T0270867

Soil: Fragile Interface

Soil
Fragile Interface

Editors

P. Stengel

Institut National de la Recherche Agronomique
Paris, France

S. Gelin

Institut National de la Recherche Agronomique
Marcy l'etoile, France

CRC Press
Taylor & Francis Group
Boca Raton   London   New York

CRC Press is an imprint of the
Taylor & Francis Group, an informa business

Reprinted 2010 by CRC Press

CRC Press
6000 Broken Sound Parkway, NW
Suite 300, Boca Raton, FL 33487
270 Madison Avenue
New York, NY 10016
2 Park Square, Milton Park
Abingdon, Oxon OX14 4RN, UK

ISBN 1-57808-219-6

SCIENCE PUBLISHERS, Inc.
Post Office Box 699
Enfield, New Hampshire 03784
United States of America

Library of Congress Cataloging-in-Publication Data

Sol, interface fragile. English
   Soil, fragile interface/editors, P. Stengel, S. Gelin.
     p. cm.
   Includes bibliographical references.
   ISBN 1-57808-219-6
    1. Soils. 2. Soil ecology. I. Stengel, P. (Pierre)
    II. Gelin, S. III. Title

S591.S781513 2003
631.4--dc21

                                    2003057302

Published by arrangement with INRA, Paris, France.

Ourage publié avec le concours du Ministère français chargé de la culture-Centre national du livre.

(This work has been published with the help of the French Ministere de la Culture-Centre national du livre.)

Translation of: *soil: interface fragile,* INRA, Paris, 1998.

French edition:© INRA, Paris, 1998
        ISBN 2-7380-0786-4
        ISSN 1144-7605

Published by Science Publishers Inc., Enfield, NH, USA

Preface

Soil: Fragile Interface presents aspects of soil as studied by pedologists and agronomists. Thus it presents the surface horizon of the Earth's crust which is subjected to climatic agents and inhabited by living creatures. These phenomena in concert slowly transform soil through the interaction of physical, climatic and biological processes. The resultant formation covers almost all continental regions and forms a pedological blanket. Depending on local climates, the rocks from which this 'blanket' is formed, the vegetation found in it, its water regime, the relief, and lastly the use to which it is put by man, it is differentiated with the formation of soils. The latter can be very varied in composition and differentiation of their constituent horizons. A description of this diversity and analysis of its genesis and the role of the aforementioned factors, help in explaining the geographical distribution of soils, their localisation in landscapes and the present changes in them. They constitute the subject matter of pedological research. It should be kept in mind that this concept of soil is not the only one. As a term commonly used by various scientific and professional communities, the word 'soil' has many meanings. The most common meaning is 'a support on which all fauna and flora move and on which all heavy structures whether living or inert are borne'. In law it is considered the equivalent of area or territory and in civil engineering the material on which structures are erected. The gardener and the agriculturist often consider soil synonymous with earth, the loose surface layer of material that is worked with a spade or a plough. Lastly, the geologist and the hydrogeologist tend to use the word soil to mean the loose surface layers, in particular those at all levels above aquifers which are not saturated with water.

The aforesaid has given rise to many ideas and images with regard to soil. The facts emphasised in this book are closely associated with the definition given above even though it is not exclusive. Soil is perceived as a major interface for the world, a blanket of the lithosphere in contact with the atmosphere. As such, it has a relationship with water from precipitations and the water tables replenished through it by drainage. The capacity of soil for infiltration and retention of water make it a favourable medium for life and an essential component of the terrestrial biosphere and an environment and habitat for most of the continental biomass. As an interface, soil is thus an ecological system whose complexity is manifest by the considerable diversity of organisms existing in it. Based on this relationship with the biosphere it is pertinent to derive an alternative definition of soil that takes into account its limit in depth. So it may be considered the vertical extension of a plant root system, which in the most frequently observed covers and for most of their bulk are found at a depth of 1 and 2 metres when no impenetrable obstacle is encountered. As a physical interface and an ecological system, these two essential aspects of soil must take into account the manner in which it is used by man, and which places it within the limits of the anthroposphere. Trophic support for agricultural prediction is obviously the first aspect associated with the concept of soil. Until recently it was a very important reason for social interest and scientific study. Even today there is serious concern because of the continuing threat to future food security given the rapid increase in human population and the eventual implications with respect to soil protection on the quality of food.

The likelihood of environmental changes has also markedly augmented concern for soil and the diversity of its functions as well as its environmental uses. As an interface, the soil also acts as a crossroad for exchanges between the atmosphere, the lithosphere, the biosphere and the terrestrial hydrosphere. It intervenes in regulation of water flow and is the place of exchange and intense biogeochemical transformations. These together enable provision of mineral elements of living matter, accumulation and purification of substances that may contaminate water and the atmosphere, as well as the recycling of organic substances and wastes produced by human life and activity.

From the aforesaid, it is evident that soil is distinctly and progressively a vital resource for mankind and the functioning of ecosystems. Concomitantly, there is greater awareness of the threats against the long-term perenniality of this resource. These threats result from many processes. In France the most worrisome are chemical pollution due to toxic trace elements or organic pollution, acidification of forest soils, erosion and physical degradation, and reduction in organic matter content. At the global level erosion and salinisation, the latter often due to irrigation with water of bad quality, and acidification have affected vast areas. Although it is presently not possible to give an accurate estimation, the extent mentioned most often is several tens of millions of hectares of arable soil per annum for a world availability of three billion hectares of cultivable soil. Degradation in the biological qualities of soil is assumed to be due to the decrease of activity and diversity of the organisms existing in it, given their progressive increment. But this remains very difficult to characterise and quantify.

Most changes in soil and its quality are only slowly reversible. Soil formation, as such, is a process whose duration comprises thousands of years. Soil ought therefore to be considered a practically non-renewable resource. So it is imperative that a way to ensure durable management of this legacy be found.

The interface is thus vital, fragile and first and foremost, non-renewable. Nevertheless, mankind, which has long associated its prosperity with the fertility of the Earth, has neglected its care for several decades. Water and air have received far more attention. This is clearly evident in France. Recent awareness of the limits to the purifying capacity of various soils with respect to nitrogen and phosphate pollution and the increasing level of pesticides in our waters, have prompted corrective measures. Such measures have been further reinforced through the realisation of the risk associated with the spread of urban wastes and the possible role of physical degradation in the genesis of catastrophic floods.

An essential element of safeguarding and broadening this corrective programme is education of students, i.e., dispelling the present naive notion of soil as a sticky mud or a place for burial of this or that with the revelation of soil as a fascinating living complex entity associated with the history of the world and mankind. This book strives to attain this objective in the hope of restoring soil to its rightful place by emphasising concern for its protection and advocating a broad-based rehabilitation programme.

Pierre Stengel
Scientific Director for Environment
Forest and Agriculture, INRA

Contents

List of Contributors

Jérôme Balesdent
DEVM-CEA
Cadarache
13108 Saint-Paul-lez-Durance

Jacques Berthelin
Centre de Pédologie biologique
UPR 6831 CNRS
54501 Vandceuvre-les-Nancy

Guilhem Bourrié
Laboratoire de Science du Sol
INRA-ENSA
35042 Rennes

Ary Bruand
Service d'étude des sols
et de la carte pédologique de
France
SESCPF
INRA
45160Ardon

Laurent Bruckler
Station de Science du Sol
INRA
Domaine Saint-Paul
84914 Avignon Cedex 9

Philippe Cambier
Laboratoire de Science du Sol
INRA
78026 Versailles

Claire Chenu
Station de Science du Sol
INRA
78026 Versailles

Claude Cheverry
Laboratoire de Science du Sol
INRA-ENSA
35042 Rennes

Cathy Clermont-Dauphin
CRDA Centre de Salagnac Damien
HAÏTI

Jean-Claude Fardeau
Laboratoire
de Science du Sol
INRA
78026 Versailles

Chantal Gascuel-Odoux
Laboratoire de Science du Sol
INRA-ENSA
35042 Rennes

Sandrine Gelin
INRA
Délégation Régionale Rhône-Alpes
1, Avenue Bourgelat
69280 Marcy L'Etoile

Jean-Claude Germon
Laboratoire de Microbiologie des
Sols
INRA
21034 Dijon Cedex

Thierry Heulin
Laboratoire d'Ecologie microbienne
de la Rhizosphère (LEMIR)
DSV-DEVM
CEA-CADARACHE
13108 Saint-Paul-Lez-Durance

Yves Le Bissonnais
Laboratoire de Science du Sol
INRA
45160 ARDON

Philippe Lemanceau
CMSE INRA
Laboratoire de Recherches
sur la Flore pathogene du Sol
21034 Dijon Cedex

Serge Martin
Service d'Etude des Sols
et de la carte pédologique de
France
SESCPF
INRA
45160 Ardon

Michel Mench
Laboratoire d'Agronomie
INRA
33883 Villenave d'Omon

Jean-Marc Meynard
Station d'Agronomie
INRA
78350 Thiverval-Grignon

Pierre Stengel
INRA
Direction Scientifique ECONAT
147, rue de l'Université
75338 Paris Cedex 07

Martine Tercé
Station de Science du Sol
INRA
78026 Versailles

Part
I

Soil: A Site of Exchange and Transfer

CHAPTER 1

Constituents and Organisation of Soil

C. Chenu, A. Bruand

Soil is the loose surface layer of the earth's crust in which plant roots develop. It is a *three-phase* medium, consisting of a solid porous phase, the pores being filled with a soil solution (liquid phase) comprising water and dissolved substances, and a gaseous phase—a mixture of nitrogen, oxygen, carbon dioxide and water vapour (Fig. 1.1). The proportion of these different phases varies with the type of soil and also over time depending on climatic conditions.

Soil is a *divided material*. The bulk of the solid phase is formed of mineral particles of variable size and of an organic fraction that is less abundant. These constituents are not arranged at random. At different levels of each, in the arrangement of solid particles in a soil profile or a landscape, there are recognisable morphological units: soil is an *organised material*.

MINERAL FRACTION: DIVIDED AND REACTIVE

The mineral fraction of a soil is the result of the fragmentation and weathering of underlying rocks. It consists of rock fragments and mineral particles of various size and properties. Classically, the mineral constituents of soils are separated according to their size; this is called

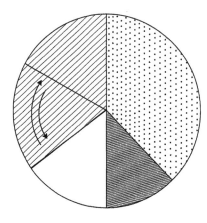

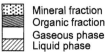

Mineral fraction
Organic fraction
Gaseous phase
Liquid phase

Fig 1 1   Relative proportion, in volume, of solid mineral and organic constituents and liquid and gaseous phases in soil.

the *granulometric analysis* of soil. The particle size of sand is between 2 mm and 0.05 mm (50 µm), of silt between—50 and 2 µm, and of clay— less than 2 µm. Their relative proportions determine the soil *texture*.

The coarse fraction of a soil, of the size of sands and silts, is called the skeleton. This skeleton consists mainly of inherited minerals: quartz, feldspars, carbonates etc. This fraction is chemically somewhat inert because the particles bear low electric charges and their specific surface area (developed surface area per unit of mass) is small. The smallest mineral constituents, i.e., the clay minerals and oxyhydroxides of iron and aluminium, are chemically the most active because of the surfaces they develop and the charge on them.

Clay

The word *clay* is defined in two ways. From the agronomical point of view clay is the mineral fraction of a soil in which particles are < 2 µm. It is, therefore, a heterogeneous fraction with respect to its nature. From the mineralogical point of view, clays are layered silicates of which the basic unit is a *sheet* with a well-defined structure. The sheets are stacked layers of tetrahedra of silica (Te) or octahedra of alumina (Oc). A sheet is made of two layers, Te-Oc, or three layers, Te-Oc-Te (Fig. 1.2). Some of the cations of this structure may be replaced by cations of a similar size, but with a lower charge: for example, substitution in the tetrahedral layer of Al^{3+} for Si^{4+}. The major consequence of these substitutions is a deficit of positive charges that causes the sheet to have an overall negative charge. This deficit may be as much as a millimole of negative charges per gram of clay (or milliequivalent per gram) (Table 1.1). This negative charge is compensated by cations on the outer faces of the sheet, which may be exchanged with cations available in the external medium: these are called *exchangeable cations*.

In a soil, the sheets of clay are not free but always associated into complex structures. The sheets are stacked in a parallel manner and separated by *interlayer spaces*. The stacking may be jointed in the case of

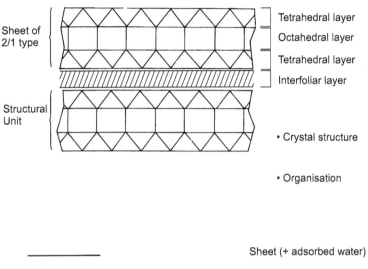

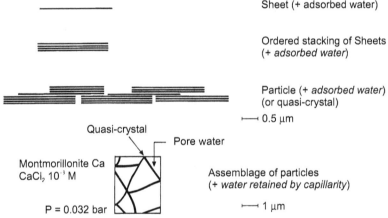

Different levels of organization in clays: example of montmorillonite (after Tessier, 1986). A sheet is the first level of organization. The second level of organization corresponds to the ordered stacking of sheets, the third to particles or quasicrystals, and the fourth to the three-dimensional assemblage of these particles.

neutral sheets or, when there is a charge on the sheet, this space is occupied by anhydrous or by hydrated cations and by water. In the case of clay minerals as in that of proteins, there may be three hierarchical levels of organisation. The sheet and its associated interlayer represent the basic unit, i.e. the primary structure. The ordered stacking of sheets represents the second structural level and the particles, which are entities that can be distinguished, a third level (Fig. 1.2). Lastly, in soils,

Table 1.1 Main clay minerals in soils. Various types of clays are identified according to type of sheet and stacking of tetrahedral (Te) or octahedral (Oe) layers and the nature and frequency of cation substitutions in the sheets

Type of stacking	Thickness of sheet	Substitution in the Te layer and extent of substitution	Substitution in the octahedral layer	Principal clays in soils				
				Name of mineral	Interfoliar cations	CEC, cmol kg^{-1}	Interstitial water	Interstitial distance
Te/Oe	7 Å	None	None	Kaolinite	None	< 0.1	None	7 Å
Te/Oc/Te	10 Å	None	Yes	Montmorillonite	Yes, various	1–1.2	Yes	Variable
		Al^{3+} 0.5	None	Beidellite	Yes, various	1–1.2	Yes	Variable
		Al^{3+} 0 < x < 1	Yes	Illite	K^+	0.2–0.3	None	10 Å
		Al^{3+} 1	Yes	Vermiculite	Various	1–1.5	Yes	Variable
		Al^{3+} > 1	Yes	Chlorite	Mg^{2+}, Al^{3+}	0.1–0.15	None	14 Å

these particles form three-dimensional assemblages, depending on the conditions of hydration, mechanical pressure and the nature and concentration of salts in the soil solution.

The small size of clay particles imparts to them colloidal properties, which were recognised in the first half of the twentieth century. Clay particles may exist in a dispersed or flocculated state, depending on the nature and concentration of cations present in the medium. The small size and form of the sheet cause clays to have a very large specific surface area: from several square metres to hundreds of square metres per gram ! Their charge makes them adsorbers and exchangers of ions and water. Water molecules are held firmly on the clay surface by adsorption. Moreover, water is also retained by capillary force in the pores formed by the assemblage of particles (Fig. 1.2). It is this water that is readily available to plants and micro-organisms.

All these properties make clays the most reactive mineral constituents, contributing considerably to the physical, chemical and biological properties of soils. Clays are also extremely sensitive to external physicochemical and physical conditions resulting from climatic changes, cultural practices such as irrigation, drainage, fertiliser application, compaction and soil amendments.

Clay minerals have been studied for several decades with respect to their various industrial applications, e.g., ceramics, construction, thinners, catalysts etc. Analysis of their crystal structure and surface properties was the focus of research until the 1980s. The microstructure of clays in relation to their physical properties was analysed thanks to developments in methods of study that ensured the preservation of their structure in the hydrated state. While studies on clays focused for a long time on those from natural beds, efforts are presently underway to characterise clays from soils. The latter are more complex and are often mixtures of different mineralogical species even within a single particle.

CONSTANTLY RENEWED ORGANIC FRACTION

When plants and organisms living in the soil die, their residues accumulate and are subjected to physical, chemical and especially biological transformations in the soil. They are biodegraded. Living organisms in the soil and organic residue at various stages of their evolution, coarse plant debris, macromolecules and single molecules constitute the *soil organic matter*. It is therefore a fraction that is constantly renewed by addition and biodegradation.

The organic fraction, on the average, represents about 5% by weight and 12% by total volume of the soil with wide variations depending on the type of soil, climate, ecosystem (forest, grassland, cultivated soil), and also on agricultural practices (e.g., removal of crop residues, tillage). Hence the organic matter content of a given soil may vary over time.

Living Organisms

Soil is home to a surprisingly wide variety of living creatures: earthworms, nematodes, insects, acarids, protozoa, micro-organisms and, of course, plant roots. There are millions, even billions, of micro-organisms per gram of soil but they represent only a small per cent of the mass of the organic fraction in soils. This microflora is taxonomically extremely diverse; it comprises bacteria, fungi, actinomycetes and algae. It is estimated that the number of species of micro-organisms in a soil exceed several thousand (Table 1.2). The soil microflora is also of highly diverse physiological and ecological types. Heterotrophic and autotrophic, aerobic and anaerobic micro-organisms coexist in the same soil profile. For this reason, soil has often been considered as an inexhaustible reservoir of different species of micro-organisms. It is not yet clear, however, whether this reservoir of biodiversity can be irreversibly affected by organic and metallic pollution in soils.

Table 1 2   *Type and abundance of soil organisms*

	Living organisms in soils	
	Number (/g soil)	*Biomass (kg/ha)*
Bacteria	10^6–10^{10}	300–3000
Actinomycetes	10^5–10^7	50–500
Fungi	10^4–10^6	500–5000 (180 m/g)
Protozoa	10^4–10^5	7–200
Algae	10^3–10^5	50–200
Fauna	10^3–10^4	500–2000 (earthworms)

The mircroflora may account for one to several tons of dry matter per hectare of surface.

Humus

The term *humus* refers to decomposed organic matter in soil, which cannot be identified visually by its form. It includes biomolecules inherited form living matter whose chemical nature can be indentified, such as polysaccharides, proteins or lignins, as well as *humic compounds*, substances that are colloidal and chemically complex.

Since the nineteenth century scientists have investigated the presence of coloured compounds in soil that are soluble in an alkaline medium and represent more than half of the organic matter in soils. Humic acids, fulvic acids and humin are fractions defined by the procedure for their extraction (Fig. 1.3). This operational definition is still used but recent investigations have shown that these humic fractions do not correspond to pools with different turnover rates nor have specific functions in the soil. The 'classical' approach of chemical fractionation using acids and bases was therefore gradually replaced by physical methods enabling direct analysis of soil organic matter without prior extraction. Concomitantly, physical methods for fractionation of organic matter, based on differences in particle size or density, were widely used because they seemed to separate fractions which, although

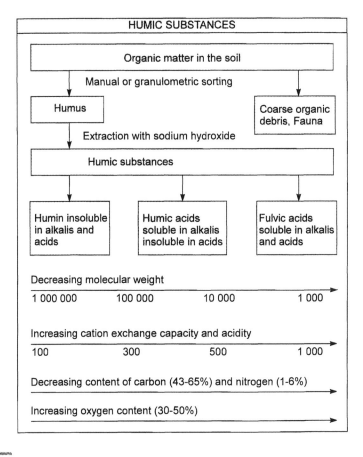

Fig 1.3   Separation and properties of humic substance.

not chemically homogeneous, serve a functional purpose, or have a given turnover rate.

What precisely are the humic substances? They are macromolecules of high molecular weight, with limited solubility in water, and amphiphilic properties. Humic substances contain a large variety of functional groups: carboxyls, hydroxyls, phenols, which confer on them a very high overall anionic charge: from 1 to 5 millimoles of negative charge per gram. They have consequently a cation exchange capacity that is greater than that of clays and may form complexes with metal. They are therefore highly reactive chemical molecules which interact in the soil with nutrients and bind to mineral surfaces as well as to organic and metallic pollutants, which explains the persistent interest of the scientific community in humic substances.

The structure of humic substances has not yet been clearly established, but considerable progress has been made with methods such as ^{13}C nuclear magnetic resonance (NMR) or pyrolysis combined with mass spectrometry. It is likely that no two humic molecules are identical in soils. Indeed biological molecules, humic substances are not a repetition of a particular sequence but are built upon elementary building blocks, more or less complex aromatic units, sugars, amino acids, hydrocarbon chains, which are randomly assembled. All these form a chain polymer arranged in a random coil, in a more or less folded state depending on the physicochemical conditions. An example is shown in Figure 1.4.

Traditionally, humus was considered the vital factor in soil fertility. It was only at the end of the nineteenth century that Liebig showed that plants do not draw their nutrients directly from organic matter but rather from mineral elements. Humus contributes to chemical fertility in soils by providing mineral elements during its biodegradation and holding cations on its negative charges. It also participates to a large extent in the physical properties in soils by its specific properties, especially water retention, and by the formation of organomineral complexes.

Organomineral Associations

During their incorporation into the soil mineral matrix, organic compounds in the soil react with the surface of minerals, in particular that of clays. Various direct bonds are established (electrostatic bond, hydrogen bond etc.) and clays and organic matter complexes (*clay-humus complex*) or oxides with organic matter are eventually formed.

Such associations are extremely frequent in soils. This is confirmed by the difficulty of separating organic matter and minerals in soils

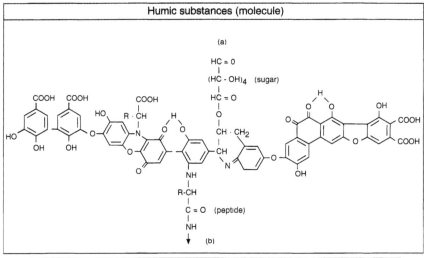

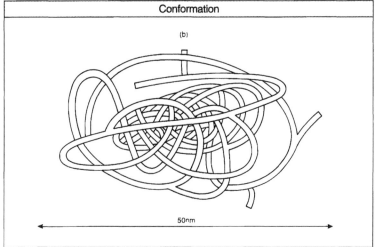

Fig. 1.4   Structure and schematic representation of humic substances (Oades, 1989).

without destroying one of these phases. Most humic substances would be involved in organomineral complexes. In fact, basic information about the formation and properties of such complexes has been obtained from laboratory studies of the complexes formed by well-known organic molecules and pure minerals, because of the complexity of natural organomineral associations. Present studies of these associations mainly involve densimetric or granulometric separation after a fairly thorough dispersion of the soil. These methods help to separate slightly decomposed plant debris from amorphous organic fractions bound to minerals of different size. Clay-humus complexes are then described with electron microscopy and X-ray diffraction as organic coatings on the surface or between the sheets of clay particles, as well as very small aggregates of fine organic debris or of micro-organisms and clays.

The formation of such complexes has very important consequences for the soil. On the one hand, the adsorbed organic matter considerably modifies the properties of mineral assemblages, especially their cohesion and/or their adsorption capacity; on the other hand, the biodegradation of organic matter adsorbed on clays is retarded.

PERPETUALLY EVOLVING STRUCTURE

Soil has an internal structure with several levels of organisation. They are either overlapping or juxtaposed and may be observed in a soil profile.

Clayey Plasma Highly Reactive with Water

The assemblage of the smallest constituents, those that make up the clay fraction according to the granulometric definition of the term (clay minerals, hydrated oxides of metals, organic molecules), forms the so-called *clay plasma*. It mainly consists of clay minerals and is amorphous when observed under an optical microscope. The geometry of this assemblage is associated with the characteristics of the clay minerals (size, form, electric charge, cations present in the interstitial space) (Fig. 1.5). At the same time, the stability of the structure is closely dependent on the constituents present in limited amounts (hydrated oxides of metals and organic macromolecules).

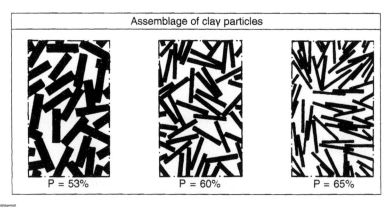

Assemblage of clay particles		
P = 53%	P = 60%	P = 65%

Fig. 5   Assemblage of clay particles in a section perpendicular to that of the sheets. This diagram, which shows that the porosity (P) of clay plasma increases with decrease in number of sheets per elementary particle, is based on studies of the water retention properties of clay soils (Bruand and Zimmer, 1992).

One of the fundamental properties of clay plasma with respect to the physical behaviour of soils is its reactivity with water, that is, its ability to retain water at the level of its hygroscopic sites (hydratable compensatory cations of clayey minerals) and especially in pores in the assemblage of clay particles. Variations in the water content of clay plasma are noted by changes in the mode of assemblage of the clayey particles and by variations in the volume of hydrated clayey plasma.

Assemblage of Skeleton and Clay Plasma with Pores

In soil, clay plasma is assembled with particles of the skeleton that are much larger than the plasma constituents (10 to 1000 times by volume). For many soils, it is accepted that the mode of assembly of skeleton particles with clay plasma depends mainly on their relative proportions (Fig. 1.6). The resultant assemblage is described by a law of mixing, the clay plasma occupying a volume that varies with the state of hydration. For this reason, soil porosity has never been correctly described by laws applicable to granular mediums. For clay contents of less than 15–20%, clay plasma mainly occupies the pores resulting from the assemblage of grains in the skeleton. When the clay content of the soil is high, the clay plasma occupies a large volume in the assemblage and pore space between the grains of the skeleton tends to be occupied by clay plasma, resulting in a decrease in the volume of the filled pores between the grains of the skeleton. For clay contents of more than 30–40%, not only is

Fig. 1.6 Skeleton-clay plasma assemblage in a silty-clayey soil in Beauce. View of a polished section of soil after embedding in a resin; scanning electron microscopy in back-scattering electron mode. Filled lacunae are shown in black, grains of the skeleton in light grey and clay plasma in dark grey.

the entire space between the grains of the skeleton occupied by clay plasma, but the grains are not in contact with each other; they are, so to speak, 'submerged' in the clay plasma. Other modes of assemblage may be observed, especially in some soils in the intertropical zone, but the model of assemblage illustrated in Figure 1.6 is the most common.

Assemblage with Variable Geometry

Variations in the volume of clay plasma, depending on the state of hydration, generate mechanical stress within the skeleton-clay plasma assemblage. These stresses are responsible for rupture planes and generate a network of cracks. Theoretically, the higher the clay content, the more the variation in volume between two states of hydration and the greater the number of cracks in the soil. In fact, this also depends on the nature of the clay minerals. In addition, depending on the clay content, variations in the volume of clay plasma are reflected to various extents at the microscopic level (microcracks within the clay plasma) and at the macroscopic level (cracks visible to the naked eye). Generally, the higher the clay content, the greater the variations in volume of clay plasma at the microscopic level. Thus for clay soils (clay content of more than 30–40%) a wide network of cracks is observed; the cracks become wider when the grid of the area is large.

Hence the porosity of the skeleton-clay plasma assemblage may vary due as much to seasonal cycles of wetting and drying as to the nature of the soil constituents, if not more so. For this reason, it is difficult to indicate the precise size of the soil pores. Not only is their geometry far from simple (e.g., a cylinder, cone, parallelepiped), but also highly variable depending on the state of hydration. Some size ranges may be given here, however: pores comprising the porosity of the clay plasma vary in size from a few microns to several nanometres. Pores corresponding to the filling lacunae between the skeleton grains have a diameter of a few microns to several tens of microns. When the clay content of the soil is more than 30–40%, the filling lacunae are isolated from one another by the clay plasma. It is possible to pass from one to the other only through the pores of the clay plasma.

Fragments Called Aggregates or Clods

After digging a trench, on examining its sides we readily discern individual fragments ranging in size from 1 millimetre to several decimetres. Actually, these are not truly fragments because they are not produced by the rupture of pre-existing objects, but definitely natural objects called *aggregates*. These aggregates, which have resulted from the combined action of cycles of wetting and drying, and soil fauna and roots, constitute the macroscopic level of soil organisation (Fig. 1.7).

The characteristics of aggregates, such as their size, structure and porosity, determine the conditions for circulation of water and gases in soil as well as its mechanical properties. Such properties are determinants for biological activities in a soil, in particular the development of roots. Morphologically speaking, there are several types

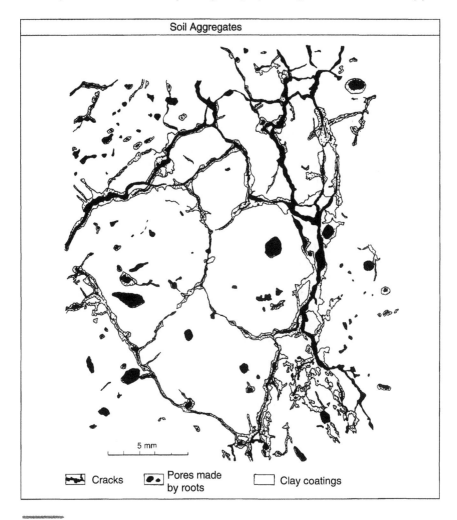

Soil Aggregates

5 mm

Cracks   Pores made by roots   Clay coatings

Fig 1 7   Sketch of soil aggregates observed in optical microscopy of a thin section. The aggregates are seen in a network of cracks with clay coatings at their periphery. A large number of nearly circular pores made by roots are also visible (Bruand and Prost, 1987).

of aggregates, each of which helps to identify the type of *structure* of the soil from which it originated. Among the most common soil structures are:

- crumb structure (porous aggregates, irregular, with curved faces);
- lamellar structure (aggregates with a preferred orientation in two directions of space and angular edges);
- angular-blocky structure (aggregates with many plane faces, with no preferred orientation and angular edges).

Analysis of soil structure was long limited to this morphological approach. For several years, attempts were made to identify the structural characteristics that mainly determine the physical behaviour of a soil. Research has also continued on the relationships between the geometry of the assemblage and its physical properties as well as on modelling soil behaviour.

In many soils the dynamics of cracking during cycles of wetting and drying is the main factor in the formation of aggregates. The activity of fauna and roots may considerably modify the characteristics of aggregates formed in this way. When soil is cultivated, the topsoil is reorganised by farm machinery and constantly modified by tilling operations that cause crumbling and compaction. Here we are dealing with structure in which the elementary units are called *clods*. Agronomists are particularly interested in the structure of the topsoil because it is the result of the combined effect of tools used for soil-working and climatic factors. A careful observation of soil helps to reveal and quantify in particular the presence of clods compacted by farm machinery.

Thin Layers Termed Horizons

Observation of a trench also reveals layers fairly parallel to each other on the soil face, of a thickness that may extend from a few centimetres to several decimetres; the *horizons*. The relative morphological homogeneity in these layers corresponds to a certain homogeneity of constitution and structure. Laterally, the horizons may extend hectometres to kilometres (Fig. 1.8). These horizons are the result of soil formation, that is, a set of transformations and displacements of mineral and organic constituents in the soil. This interpretation of horizons is fundamental for establishing the topology of soils and linking them with a reference system or classification.

Based on their characteristics, the horizons may be divided into three major groups from the surface downwards:

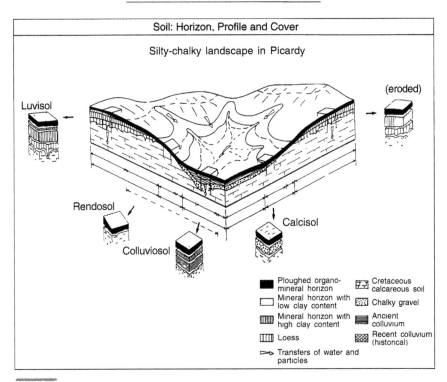

Fig 1 8   Block diagram (a landscape from Picardy) illustrating the concepts of horizon, soil profile and soil cover (Begon and Jamagne, 1994).

- holorganic (entirely organic) horizons, which essentially comprise the debris of more or less transformed, but still recognisable plants;
- organomineral horizon, differentiated from the holorganic horizon by the presence of moistened organic compounds and minerals constituents in a large quantity;
- mineral horizons, where the assemblage of constituents is reorganised and geochemical weathering of constituents occurs by removal or accumulation of elements (migration of clays, chemical elements such as iron or aluminium);
- lastly, horizons that correspond to the underlying rock, subjected to some weathering and disintegration.

Soil Profile and Cultural Profile

The soil profile corresponds to the superposition of all the soil horizons, of which several elements have been schematically presented in Figure

1.8. All the horizons distributed in space constitute the *soil cover*. The vertical succession of horizons as well as their distribution in the third dimension is generally the result of processes of soil formation that may be very old, especially in the intertropical zone. In Western Europe, the processes of soil formation from which the present soils originated often developed only after the last major Ice Age, which occurred about 10,000 years ago. These processes of soil formation are slow and modify the soil over long periods of time.

Within a horizon, most of the skeleton-clay plasma is assembled according to a basic pattern, the characteristics of which have already been mentioned. Locally reorganisations and concentrations (clayey coatings on the faces of pores, concretions) called *pedological features* are observed. These features or traits, closely associated with the processes of soil formation, provide important information for understanding the manner in which a soil is differentiated over time.

The topsoil of land used for agricultural production, i.e., the horizons inherited from soil genesis, changes within several decimetres and in a short period. The natural organisation of the soil horizons is so modified that the pedogenic soil features almost disappear or are displaced within the soil (Fig. 1.9). It is in the ploughed (tilled) layer that the structure is most strongly modified. Only the skeleton-clay plasma assemblage does not change much because it is essentially determined by the nature and proportion of soil constituents. On the other hand, at the base of the tilled layer and one or two metres deep, there is often a superposition of organisations due to cultural practices and those of soil

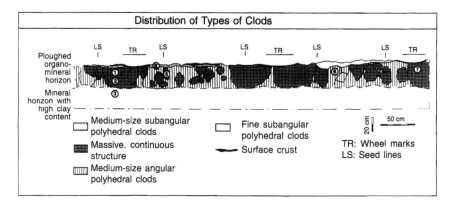

Fig. 1.9   Distribution of different types of clods in a cultivated horizon. Note the presence of zones of large size to the right of recent wheel marks. The subangular polyhedral and fine structure present at the surface is due to soil preparation for sowing seeds (Bruand et al., 1993).

origin. Deep down, the natural organisation of the soil is not altered, except when certain ameliorative works have been undertaken (e.g., drainage, subsoiling).

Macroscopic organisations associated with the effects of farm tools and climatic factors define the *cultural profile*. Analysis of this profile provides a wealth of information to the agronomist who would like to estimate the effects of different tools used during soil-working and the stability of the structural elements thus formed when they are subjected to climatic factors.

CONCLUSION

On a human time scale, the mineral fraction of a given soil may be considered to be a constant composition because the pedogenetic transformations that form it occur over thousands of years. The same does not apply to the organic fraction, which is quantitatively and qualitatively affected by land use and agricultural practices and therefore soil structure depends, as already mentioned, not only on the nature and proportion of the various constituents, but also on external physical and physicochemical conditions resulting from climate and human activity, such as tillage and irrigation. Therefore, organisation of the constituents that determine the geometry and volume of soil porosity and consequently the conditions for retention and circulation of water, solutes and air, may vary over a human time scale.

Glossary

Aggregate	A natural unit of a coherent assemblage formed of elementary soil particles.
Clay	1. A mineral fraction of soil in which the particles are < 2 μm.
	2. Layered silicate minerals.
Clay plasma	An assemblage of constituents in clay fraction of soil (< 2μm).
Clod	A compact and cohesive unit of soil with a diameter of a few millimetres to several centimetres, produced in soil by human activity such as ploughing.
Cultural profile	Superposition of macroscopic organisations in cultivated soil, formed by the action of tools, growth of roots and climatic agents, observed in a vertical section.
Fulvic acid	Organic compounds in soils or surface waters, soluble in alkalis and acids.
Horizon	Layer roughly parallel to the soil surface. Horizons differ from each other in their constituents and organisation. They have differentiated during the course of soil formation.

Humic acid	Organic compounds in soils or surface waters, extracted by dissolving in alkalis and precipitation in acids.
Humus	1. Decomposed organic matter of soil with a dark colour.
	2. Surficial organic part of forest soil.
Organic matter	Living organisms in soil and organic residues at different stages of their formation, coarse plant debris, macromolecules and single molecules.
Skeleton	Coarse mineral particles in soil (> 2 μm).
Soil cover	Distribution of soil horizons in space (metres to kilometres).
Soil profile	Superposition of pedological horizons in a soil that can be observed in a vertical section.

∫ URTHER READING

Andreux F, Mugnier Lamy C. 1994. Genèse et propriétés de molécules humiques. In: Constituants et propriétés du sol. M. Bonneau, B. Souchier (eds.). Masson, Paris, pp. 109-142.

Begon J-C, Jamagne M. 1994. Genèse, typologie et utilisation des sols. Techniques Agricoles, 1110 (3-994): 1-24.

Bruand A, Prost R. 1987. Effect of water content on the fabric of a soil material: an experimental approach. J. Soil Sci. 38: 461-472.

Bruand A, Zimmer D. 1992. Relation entre la capacité d'échange cationique et le volume poral dans les sols argileux: incidences sur la morphologie de la phase argileuse à l'échelle des assemblages élémentaires. C. R. Acad. Sci. Paris, ser. II, vol. 315, pp. 223-229.

Bruand A, Chenu C. 1995. Constitution physique du sol. Tech. Agric. 1130: 1-12.

Bruand A, D'Acqui IP, Nyamugafata P, Darthout R, Ristori GG. 1993. Analysis of porosity in a tilled "crusting soil" in Zimbabwe. Geoderma, 59: 235-248.

Bruckert S. 1994. Analyse des complexes organominéraux des sols. In: Constituants propriétés du sol. M. Bonneau, B. Souchier (eds.). Masson, Paris, pp. 275-296.

Jamagne M. 1994. La cartographie des sols. Analyse spatiale de la couverture pédologique. In: Constituants et propriétés du sol. M. Bonneau, B. Souchier (eds.). Masson, Paris, pp. 587-618.

Oades J.M. 1989. An introduction to organic matter in mineral soils. In: Minerals in Soil Environments. D.E. Dixon (ed.). Soil Sci. Soc. Amer., Madison, WI, pp. 89-158.

Pédro G. 1994. Les minéraux argileux. In: Constituants et propriétés du sol. M. Bonneau, B. Souchier (eds.). Masson, Paris, pp. 47-64.

Tessier D. 1984. Étude expérimentale de l'organisation des matériaux argileux. Hydratation, gonflement et structuration au cours de la desiccation et de la réhumectation. Doctorat d'Etat, Univ. Paris VII, Éditions INRA, 361 pp.

CHAPTER 2

Reactivity of Soil: Chemical Properties

Martine Tercé

SOIL, A CHEMICAL REACTOR

Advancements in Knowledge

It was only during the 19th century that experiments revealed the 'absorbing power' of soil, which is its ability to fix nutritive mineral elements and release others. Prior to this, the cleansing power of soil with respect to putrefiable matter has been known from antiquity but the fundamental aspects of knowledge regarding chemical reactions in soil were based on plant nutrition. For a long time our ancestors believed that plants obtained their food from fats in manure or amendments spread on the soil. In the 16th century, the scientist Bernard Palissy was the first to state that plants obtained their food from manure spread in plantations. By the end of the 17th century, scientists had accepted the idea that plants are nourished by substances they extract from the soil.

In the middle of the 18th century, the Swedish scientist Wallerius hypothesised that, in soil, humus alone was the reserve of nutritive elements for plants. At the same time, Wallerius recommended a study of the chemical composition of soils. The chemical fertiliser industry developed during the same period. Until the 20th century, researchers did not explain the chemical process that occurred in soils which was responsible for absorption and transformation of mineral fertilisers. It was only after the Second World War that our knowledge of chemical reactions in soils greatly increased. The reasons for research conducted on this subject were initially purely agronomical. They are now directed more towards the chemical quality of soils and the role of chemical reactions due to the dispersion of pollutants in the environment.

Principal Factors: Soil Constituents

The solid, liquid and gaseous phases in soil interact and give rise to many chemical reactions. The different interfaces—solid-liquid, solid-gas, solid-solid, and liquid-gas—are the usual sites of these reactions.

The smallest fraction of soil (less than 2 µm) is the most reactive solid phase (Chapter 1). It comprises crystalline mineral compounds such as clays, oxides, hydroxides, carbonates and amorphous organic compounds (humic substances), and also minerals. These compounds are found either as isolated particles or as mineral-mineral or organomineral associations. The particles have the properties of colloidal substances, such as that of forming dispersed phases by being finely dispersed in an aqueous medium, or even aggregating, depending on the chemical composition of the soil solution.

> In 1881, the chemist Graham was the first to recognise the particular properties of colloids and to study them. In 1888, the chemist Van Bemmelen showed that it is the soil colloids that retain soluble mineral elements useful to plants.

The liquid phase is also called the soil solution because it contains dissolved compounds naturally present or added by man. These are various mineral ions formed by dissolution of soil constituents or application of fertilisers or amendments. Besides the mineral elements essential for plant life, it also contains organic and mineral compounds resulting from the decomposition of organic matter of plant origin and from the biomass (bacteria, fungi, algae, soil fauna) or pesticide products.

The gaseous phase comprises nitrogen, carbon dioxide and sometimes methane and gaseous pesticides.

External factors also influence the chemical reactivity of soils. These are temperature, precipitation, and cultural practices.

The temperature regulates the intensity of chemical reactions with respect either to their rate or amplitude. Precipitation determines the reactions at different interfaces. Cultural practices also determine the extent of inputs and removal of mineral and organic compounds and water, retention of which may be regulated by irrigation or drainage.

DIVERSITY OF CHEMICAL REACTIONS

Chemical reactions in soils are of various types: adsorption-desorption, precipitation-dissolution, oxidation-reduction (redox), acidic-basic, complexa-tion.

These reactions may occur independently or simultaneously.

The rates of such reactions are highly variable: adsorption-desorption reactions may take a few minutes to several days while precipitation-dissolution reactions may continue for decades or even centuries.

The types of reactions and their extents are variable in space due to soil heterogeneity, which itself is variable in time (in composition, water content, porosity etc.).

Reactions at the Solid-liquid Interface

The most well-studied reactions are those that occur at the solid-liquid interface: these are the reactions of adsorption and desorption, precipitation and dissolution. All these reactions, being more or less reversible, provide the soil solution with nutrients (major and trace elements) or elements that are toxic to plants (some metallic trace elements or aluminium) and are responsible for the absorptive capability of the soil, as defined during the 19th century.

Adsorption and desorption reactions

Many soil scientists make a distinction between the phenomenon of adsorption and ion exchange. This distinction is more historical than scientific, as will be shown later.

> *Adsorption* is a term that was defined in 1881 by the chemist H. Kayser. Earlier, chemists used the term *absorption* for reactions in which they noted the disappearance in a solution of chemical substances when the solution came into contact with a solid phase. Soil chemists used both terms in a somewhat confused manner until the 1930s.

Colloids in soils have what are called surface charges localised on the surface of particles. The charge on clays is usually permanent and negative, whatever the pH of the medium, and results from ion substitutions within crystallites (Fig. 2.1). The charge on oxides and

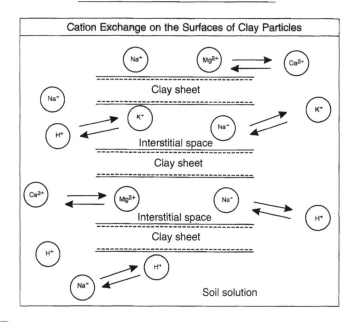

Fig 2.3   Cation exchange on the surfaces of clay particles.

The negative charges on clay sheets are compensated by the presence of cations held on the surface by ion combinations. These compensatory cations are exchangeable with cations in the soil solution.

hydroxides is variable depending on this acidity and may be positive, neutral, or negative, depending on the amphoteric nature of the OH groups on the surface (Fig. 2.2). The charge on humic substances is always negative but of variable intensity depending on the pH of the medium and is due to the dissociation of functional acid groups (mainly carboxylic). These electric charges are compensated by ions adsorbed on the surface that are exchangeable with other ions in the soil solution. This property of soil colloids was the prime mover in the earlier work on one of the main mechanisms of adsorption that occurs in soils: adsorption by ion exchange. The exchange of ions introduced on soils by mineral fertilisers or amendments with those of soil colloids was the subject of many studies until the 1960s, the objective being productive agriculture. Efforts were devoted mainly towards cation exchange of major ions such as potassium, ammonium and calcium by clays, humic substances and the clay-humus complex. Anion exchange was not studied in detail except in the case of phosphate exchange by oxides.

Research on adsorption of organic molecules (single or complex) or of living organisms (viruses, bacteria) by clays increased after 1965 and revealed that there were adsorption mechanisms other than ion exchange. These mechanisms are due not only to the presence of a surface

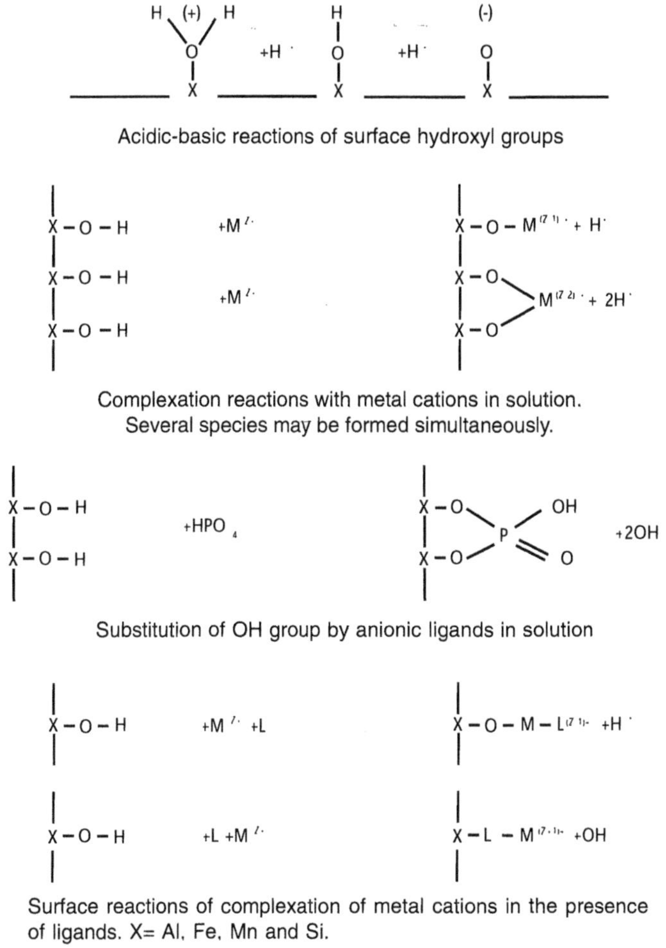

Acidic-basic reactions of surface hydroxyl groups

Complexation reactions with metal cations in solution.
Several species may be formed simultaneously.

Substitution of OH group by anionic ligands in solution

Surface reactions of complexation of metal cations in the presence
of ligands. X= Al, Fe, Mn and Si.

Fig. 2.2 Some examples of co-ordination reactions at the oxide-soil solution
interface.

electric field, but also to the crystal structure of mineral colloids or to the
nature of functional groups on the surface of mineral or organic colloids.

Ions or adsorbed molecules may revert in solution in the liquid phase:
this is the phenomenon of **desorption**, which depends on the nature of the
forces acting between the surface of the solid phase and the solute.

The intensity of adsorption and desorption depends on conditions in
the medium in which they occur and ionic composition of the soil
solution, nature of the adsorbent and adsorbate, water content and
temperature.

After the 1970s, the appearance of many pesticide type synthetic organic molecules as a result of great strides made by the pharmaceutical industry, encouraged their progressively more frequent use in agriculture. Records of the conformation of these molecules should now provide data on the mobility or persistence of these molecules in soils since these phenomena involve potential risks for the environment, water and the food chain. This has led to studies on adsorption of these molecules by soils and their constituents.

After the first oil crisis in 1973, sludge from waste treatment plants was used for agricultural purposes to reduce imports of industrial fertilisers, especially phosphate fertilisers. This, however, began to raise problems with regard to the future state of the traces of metal cations from these sludges in soils. The study of adsorption of cations by clays and humic substances received a fresh impetus.

Nevertheless, as opposed to fertilising elements, heavy metals, like pesticides, are found in soils at very low concentrations. A study of their adsorption needed improved methodologies at the theoretical and experimental level.

Also during the 1970s, there were further studies on ion adsorption by mineral oxides and hydroxides. These efforts were undertaken following agricultural developments in tropical and subtropical regions where these compounds are predominantly used. For a long time these minerals were considered inherently much less reactive than clays because they do not develop as large a surface or negative electric charge. In fact, they do not have the same interfacial properties as their electric charges vary with the pH of the medium. They are able to more easily adsorb trace metal cations by mechanisms other than that of ion exchange.

Studies conducted on adsorption of pesticides and heavy metals have constituted a turning point in agronomy from the production-oriented objectives of the 1950s to the present environmental objectives.

Precipitation-dissolution relations

Knowledge of the genesis and weathering of mineral constituents has resulted in research efforts undertaken on these reactions. Chemical weathering of minerals takes place mainly due to hydrolysis, causing elimination of many ions and formation of colloidal precipitates. These studies, which have made great strides since the 1960s, have significantly increased our knowledge of soil science.

The role of dissolution reactions in the transformation of minerals was highlighted at the beginning of the 20th century after microscopic observations on soil samples.

The dissolution of minerals through, for example calcite ($CaCO_3$), was studied in detail owing to interest in it from the agronomical and physiological points of view. It causes release of calcium ions that are fairly well tolerated depending on whether the plants are calcicoles or calcifuges, and the formation of carbon dioxide which helps to fix the pH because of the buffer effect of the soil.

The enigma of atrazine

Atrazine is a herbicide molecule used extensively in agriculture since the 1970s for freeing plantations of weeds, especially in maize, but is also used, for example, by the French Railways for maintaining the tracks. It is a reliable molecule at the plant sanitation level but not highly biodegradable and is now found in stagnant waters and some water courses. Since the time it was synthesised researchers have been working on what happens to it in soils but have not yet clarified all the mechanisms of its retention in soils. When these studies began, researchers were of the view that clays reactive with organic molecules soluble in water were the main factors in atrazine adsorption. Experiments revealed that this was not so, as the molecule is hydrophobic. Researchers then directed their attention to the reactivity of such humic substances as were more reactive than clays. The state of knowledge with respect to humic substances was not sufficiently advanced, however, for them to understand completely the mechanisms of adsorption and desorption of atrazine in soil. Many questions still remain to be solved: is atrazine adsorbed or only trapped within the pores of the three-dimensional structure of humic substances? What part is played by non-humic organic substances in the dispersion of atrazine in the environment? In spite of the extensive use of this product many questions remain, and many lines of research are still to be pursued at present.

Studies on what happens to heavy metals in soils resulted in a revival of interest in these reactions. Metal cations in traces may be precipitated in the form of a hydroxide depending on the pH of the soil solution and be temporarily immobilised depending on its variations.

Reactions at the Solid-Gas Interface

Reactions at the solid-gas interface did not evoke the same interest as at the solid-liquid interface, because the gaseous phase of a soil contains a smaller number of compounds than the liquid phase and depends on soil porosity.

Adsorption of water in the vapour state by clays was further studied during the 1970s to obtain a better understanding of the elementary phenomena involved in the transfer of solutions in soils.

Apart from water vapour, on which much work was done, the use of gaseous nematicides in agriculture was also responsible for research on adsorption of these products, for example, on soil constituents.

Reaction at the Solid-Solid Interface

Interactions between colloids, which are usual rather than exceptional, have rarely been shown in the solid state because of the difficulties in recreating experimental conditions similar to the ones observed in nature. This is why, for example, a study of clay-humus associations is conducted using humic substances extracted from the soil and placed in suspension. Obviously, these experimental conditions do not adequately represent the ones that prevail in nature.

Reactions at the Liquid-Gas Interface

Reactions at the liquid-gas interface are few and, essentially, deal with a study of the dissolution of carbon dioxide in the soil solution resulting in bicarbonate and carbonate ions. The solubility of carbonate depends on the partial pressure of CO_2 in the air in equilibrium with the solution, temperature and concentration of the dissolved electrolytes in the soil solution. Carbonate ions help to regulate the pH of the soil solution, which is a buffered medium, a condition favourable for root development.

Reactions in Liquid Phase

Reactions in the soil solution are of three types: acidic-basic, redox and complexation.

Acidic-basic reactions are due to the dissociation of mineral and organic acids found in the soil solution.

Redox reactions are mainly of biochemical origin and rarely chemical. Oxidation of iron in manganese-rich soils is, however, an exclusively chemical process.

Studies on complexation reactions of toxic metals in the form of traces and pesticides by water-soluble organic matter in soils are recent and are now being conducted for a better understanding of the dispersal of these products in the environment. Interactions in solution of trace metals and pesticides are of considerable importance because they maintain them in solution whatever the conditions in the medium. Figure 2.3 summarises the different reactions of heavy metals in soils.

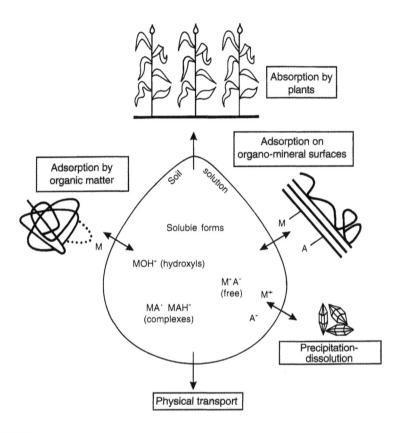

Fig. 2.3 Reactivity of soil constituents to heavy metals.

TOOLS FOR THE STUDY OF CHEMICAL REACTIVITY IN SOILS

Experimental Tools

X-ray diffraction has been used since 1936 to examine the structure of crystallised mineral colloids. Since the 1960s, infrared absorption spectroscopy has been used to study adsorption of water in the vapour state, or organic molecules that are highly soluble in water. These methods have not been found suitable for study of the adsorption of trace elements or molecules. Electrochemical and electrokinetic methods were introduced in the 1970s to study the phenomena at solid-liquid interfaces and in solution.

Improvements in electron microscopy and its induction in soil science helped in the development of knowledge with respect to the processes of precipitation and dissolution that take place in soils.

The progress in analytical chemistry as applied partly to the analysis of water has improved our knowledge of the reactions involving pesticides and heavy metals in solution and at the solid-liquid interfaces.

Theoretical Tools

Adsorption of cations of fertiliser elements was quantified at first by using the laws governing ion exchange. Subsequently more elaborate theories explaining the effect of an electric field on the distribution of ions at the interface were applied to clay-solution systems about thirty years ago. These theories were not commonly applied, however, and moreover dealt with constituents modelled on the soil.

Description of ion adsorption by mineral oxides or hydroxides resulted during the 1980s in new mathematical models based on co-ordination chemistry of solution and interface. Some models even took into account the precipitation-dissolution reactions that occurred concurrently with adsorption, as in the case of heavy metals. Attempts are presently underway to generalise these new models of adsorption with other soil constituents so as to apply them to soils in their entirety.

On the other hand, a description of adsorption of pesticides and their constituents is still based on empirical models, often on theories of gas adsorption.

Glossary

Absorption	Penetration of elements in a body.
Absorption capability	Ability of soil to retain elements.
Adsorbate	A chemical element with an affinity for the surface of an adsorbent.
Adsorbent	Colloidal compound with a surface chemical reactivity.
Adsorption	Fixation of an element on the surface of colloids.
Buffer effect	Ability of a soil to resist a change in acidity due to acids or bases.
Colloid	A compound of very small size developing a large surface.
Desorption	Reversion in solution of an element fixed on the surface of a colloid.
Herbicides	Molecules intended to destroy weeds.
Ligand	A mineral or organic molecule that can combine with another molecule.

Nematicides	Molecules intended to destroy nematodes.
Pesticides	A generic term to describe all toxic molecules meant to control pests.
Purification power (cleansing power)	Ability of soil to eliminate decaying substances.

3

Transport in Soil

Laurent Bruckler

SOIL, A POROUS PERMEABLE MEDIUM

Soil is composed of three phases: a solid phase of 'particles' (sands, silts, clays), a liquid phase of water or, more precisely, the *'soil solution'*, that is, water, mineral and dissolved organic substances, and a gaseous phase of a mixture of gases (nitrogen, oxygen, carbon dioxide, water vapour etc.). *Soil porosity* includes all that is not solid phase: it thus corresponds to the total volume of liquid and gaseous phases. This porosity generally represents 30 to 60% of the total volume. Thus, about half the soil volume is 'full of voids', which are spaces filled with gas or water. Lastly, soil is a highly porous medium (Fig. 3.1); it is permeable and a likely centre of phenomena for transport of various kinds (water, dissolved elements, gas, heat). If soil were made only of solid matter, it would have been able to transport heat by conduction (like, for example, a metal rod) but it would not have allowed water to penetrate it... .

Further, the gas or water-filled spaces are in contact with *solid surfaces*. As a result, during its transport the soil solution may be filled with mineral or organic substances, or could, on the other hand, release these constituents to the solid phase. Thus, the transport of mineral or organic elements in soil is not generally passive but usually a phenomenon in close interaction with the solid phase of the soil or the

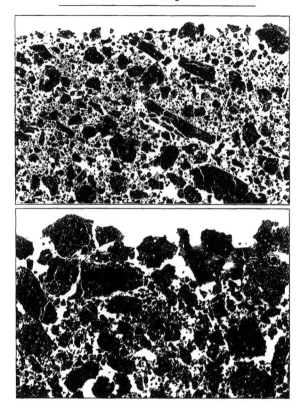

Fig 3.1   Sections of cultivated soils showing the high level of soil porosity (black parts correspond to fragments of earth, white parts to pore spaces) (Scale 1.4).
Plate : J.-C. Fiès.

micro-organisms that are part of it. There is therefore transport and physicochemical modification of what is being transported.

In all, soil, due to its porosity and constitution, is a medium suitable for the transport of liquids, gases and heat. With respect to liquids, it functions as a highly interactive medium playing, for example, the role of a purifying (cleansing) filter or conversely, charging the liquid phase with organic and mineral constituents acquired during its passage.

(ALMOST) NOTHING MOVES AND (ALMOST) NOTHING IS VISIBLE IN THE SOIL

Diversity of Transport Phenomena in Soils

Under natural conditions the fluids in soils are constantly on the move. We speak of cycles (water cycle, nitrogen cycle) which involve simultaneous processes of transport and transformation. A few examples are given here.

The soil is warmed during the day and cools down at night due to transport of heat (Fig. 3.2). Near the surface the variation between day and night temperatures may be as much as several tens of degrees. Micro-organisms in the soil, the microflora and microfauna, may thus be subjected to considerable temperature fluctuations during the course of a day, or a season merely due to heat conduction in the soil.

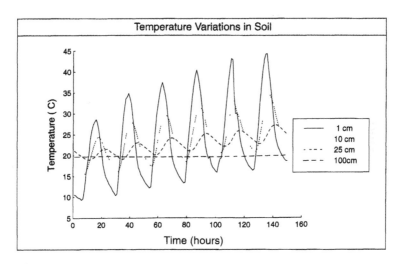

Fig. 3.? Variations in soil temperature at various depths in summer. This figure shows the considerable temperature variation and phase difference of temperature with depth.

Evapotranspiration from the plant cover (evapotranspiration from the soil is equal to the sum of direct evaporation of water removed from the soil and transpiration from the plants, that is, the water taken from the soil by roots and that which evaporates in the atmosphere through the leaves) is a continuous and determining phenomenon in the water cycle. A wheat farm in the Paris basin may lose 5000 m^3 of water per hectare from sowing to harvest, i.e., over a period of several months. This water, earlier circulated in the soil by infiltration (during rainshowers), moves to a point in the soil near the roots which absorb it. Generally, most of the rainwater that reaches the soil (usually more than 90%) is recycled by evapotranspiration. Together, all these water movements induce permanent variations in the soil water content (Fig. 3.3).

During heavy rains in autumn, spring and winter in the climate of Western Europe, water may run off the surface or percolate to a depth in

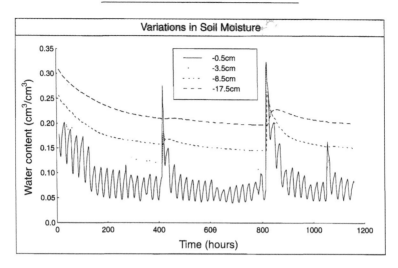

Fig. 3.3   Variation in moisture content of a soil at various depths. Decreases in moisture are due to evaporation and increases are the result of rainshowers.

the soil and may also be drained off and replenish the water tables at depth. This same water can be used in summer as potable water when rains are infrequent. The example illustrates the *regulatory role* of soil in relation to the climatic phenomena which, by their nature, are difficult to forecast, random and irregular. This replenishment of water tables during the rainy season also has adverse effects because the drained water may, for example, carry highly mobile nitrate into the soil with the water as it moves to depths. These properties are the basis of the present problems of *pollution of aquifers* by nitrate, especially in places where intensive agriculture using mineral fertilisers, or organic manures (litter) and irrigation, is practised.

Life of the nitrate ion

The case of the nitrate ion is typical of the complexity in processes of transport and transformation in the natural medium. Consider a nitrate ion brought to an agricultural plot in the form of a synthetic fertiliser made from gaseous molecular nitrogen. This nitrate ion may be 'consumed' by the micro-organisms in the soil and be immobilised temporarily into an organic molecule (for example, a protein). When the micro-organisms die, this protein is decomposed and the nitrogen may eventually return to its mineral nitrate form. After a heavy rainstorm, this ion will be transported to a depth towards the aquifer. In this zone, poor in oxygen, the

---

Contd.

> phenomenon of *denitrification* will then transform the nitrate ion to gaseous molecular nitrogen. This gaseous nitrogen may then *escape* through the soil and thus return to the atmosphere from where it came; the cycle is completed... .

These 'perpetual movements' are even more true in the case of gases, which are highly mobile and move constantly between the soil and the atmosphere. Soil micro-organisms that colonise mainly *surface horizons* of a soil and the roots of plants, are living creatures which consume oxygen for their metabolism and, in turn, produce carbon dioxide. Oxygen from the air must be transported through the soil to sites where it is consumed and, conversely, the released carbon dioxide must be removed to avoid asphyxiation of the soil. This phenomenon of asphyxiation of soil arises when the soil is unable to let gases pass through it (a too compact soil with low porosity and waterlogged) and may result in catastrophic agronomical consequences: lack of germination of seeds, death of plants not resistant to asphyxia etc. Without renewal of gases from soils, plant life would be impossible: there would be no germination, no growth and no decomposition of organic matter.

Thus soil is a dynamic entity, constantly subjected to external effects (heating due to sunlight, rain, evaporation etc.) which result in permanent exchanges, especially between the soil and the atmosphere, the soil and living organisms (micro-organisms, roots etc.).

Principal Mechanisms Involved

Although many elements in the soil are mobile, the major types of mechanisms involved are fewer and often identical for a large number of transported elements.

We may disregard the process of transport of heat which corresponds to a *transfer of energy* in soil and not a transfer of matter. The principal process that accounts for the transport of heat in soil is *conduction,* in a manner similar to heat conduction (fast) in a metal rod or (slow) in an insulated medium. This is a phenomenon of transfer of heat from one point to another through a conductive element.

In the case of water, we can identify two major types of mechanisms: the flow of water by gravity (i.e., subject to the effect of its own weight), which, for example, occurs during rapid flow of water along *'preferred paths'* in the soil (fairly large cracks or holes of earthworms through which water circulates rapidly during a heavy rainshower), and the slower transport of water in soil from one nearby point to another by diffusion through smaller soil pores (of micron size). This latter mechanism is the one most frequently observed and describes most types of transfers of water in a soil.

For solutes and gases there are two important mechanisms which account for most of the movements. On the one hand is the process of *convection* (from the Latin *cum* which signifies *with*, and *vector* which has the same English meaning) that corresponds to transport of a particular element with the movement of an entire mass. For example, in the case of the nitrate ion, NO_3^-, this will involve transport of nitrate dissolved in water with the fluid (water) that moves in the soil. In the case of carbon dioxide it will be the movements of CO_2 molecules with the entire gaseous phase containing several gases (nitrogen, oxygen, water vapour), that moves from one point to another in the soil following a gradient in total pressure between two points. On the other hand, the process of *diffusion* begins even when the entire mass of the fluid is at rest, but when the concentration of a particular body in it is not uniform from one point to another. Let us take the example of a nitrate. If nitrate ions are dissolved in an immobile liquid phase they may move spontaneously from one point to another in the phase while at rest, if the concentration of nitrates is not the same at each point. Similarly, even if the gaseous phase in the soil is immobile (and, therefore, at constant total pressure), molecules of carbon dioxide may move from one point to another, if the partial pressures of gases are not the same at each point (for example, they will be higher near a root that produces carbon dioxide than near a zone with no roots). Lastly, these transport processes generally result in a phenomenon of *dispersion* in the medium, which signifies that a highly localised provision of a particular element (water, solute, gas) will eventually spread over a progressively wider space with time. Dilution and dispersion in a medium likewise generally follow the phenomena of transport in porous mediums.

By and large, conduction (of heat), convection and diffusion explain the transfer of matter and energy in soil. Table 3.1 gives the basic laws of conduction or diffusion for the principal types of transport involved (heat, water, gases, solutes). It can be seen that the fundamental laws always have the same form: the flow is equal to the product of a coefficient of transfer that characterises the medium, and a gradient (of concentration, pressure or potential).

In addition, when we consider the transport of matter or energy in a soil, two properties are always involved regardless of the transported element:

- Direction of flow is determined physically and unambiguously: heat always moves from warmer to colder surfaces, gases or elements in solution always move from high-pressure to low-pressure areas or from high concentrations to low concentrations, water always moves from a higher to a lower level during gravitational flows, from higher to lower pressures for flows under pressure and, generally, from humid to dry mediums in the

Table 3 ▪ *Basic expression of principal flows in soils (heat, water, gases, solutes). The formal analogy between various types of flows is indicated.*

Type of transfer	Flows	Principal flows in soils	
		Coefficient of transfer	'Motor' gradient
Heat	$q_c = -\lambda \dfrac{\partial T}{\partial z}$	Thermal conductivity (λ)	Temperature $\dfrac{\partial T}{\partial z}$
Water	$q_v = -K(\theta)\dfrac{\partial \psi}{\partial z}$	Hydraulic conductivity ($K(\theta)$)	water potential $\dfrac{\partial \psi}{\partial z}$
Gases	$q_q = -D_g \dfrac{\partial C_g}{\partial z}$	Coefficient of gaseous diffusion (D_g)	Gas concentration $\dfrac{\partial C_g}{\partial z}$
Solutes	$q_s = -D_s \dfrac{\partial C_s}{\partial z}$	Coefficient of liquid diffusion (D_s)	Solute concentration $\dfrac{\partial C_s}{\partial z}$

case of flow by diffusion in soil. All these movements thus tend to *homogenise the medium* (in terms of temperature, pressure and concentration) with a tendency towards a more stable thermodynamic state.

- All soils do not have the same ability to store matter or energy. This is one of the important criteria for differentiating soils with respect to each other. Thus, wet sand is an effective filter and a good conductor of water while a clay soil is less conductive and less capable of conducting water (Table 3.2). Soils are thus differentiated simultaneously by their storage properties and their varying ability to transport a particular element. Roughly speaking, it may be stated that a sandy soil contains a low water reserve but ensures a rapid flow, while on the other hand, a clay soil has a high water reserve but ensures only a limited flow, and a silty soil is intermediate between these two.

ANALYTICAL METHODS

Several methods exist for qualifying all these phenomena of transport that affect both agricultural production (movement of water or nutrients in soil and nutrition of plants) and the quality of our environment (pollution of aquifers, salinisation of soils). We shall describe two major methodological groups.

Mass balance

As the name indicates, this involves a balance of the gains, losses and variations in the stock of a given volume of soil, written algebraically to show that the mass is conserved and therefore nothing is lost and nothing is created; everything is transformed (see inset). For example, in the case of water (Fig. 3.4), analysis of Equation (II) reveals that three terms must be known or calculated to quantify deep drainage: total rainfall and irrigation, variation in water reserve during the period under consideration, and actual cumulative evapotranspiration. With respect to total rainfall, a knowledge of this on the scale of a plot (local scale) or on several plots (regional scale) supposes direct measurement (rain gauge) or reference to meteorological data. Irrigations must necessarily be recorded by the farmer using direct measurements on plots (rain gauge) or by inquiry.

Knowledge of the variation in water reserve may be obtained by a detailed follow-up of a particular plot but this is now less frequent in observational networks because it necessitates regular measurements of the water content in a soil. Lastly, the case of actual evapotranspiration is

Table 3.2 *Extents of hydraulic conductivity at saturation between several soils (application to agricultural drainage). A very wide range of variation is noted between sandy and clayey soils (after Musy and Soutter, 1991)*

	Hydraulic conductivity										
Ks (m/s)	10^{-1}	10^{-2}	10^{-3}	10^{-4}	10^{-5}	10^{-6}	10^{-7}	10^{-8}	10^{-9}	10^{-10}	10^{-11}
or about			100 m/day	10 m/day		1 m/day	0.1 m/day				
Permeability	Permeable					Semipermeable			Impermeable		
Types of soil	Gravel without sand and fine elements			Sand with gravel. Coarse to fine sand		Peat	Very fine sand. Coarse silt to clayey silt			Silty clay to homogeneous clay	
Drainage possibilities	Excellent			Good				Moderate to poor		Poor to nil	

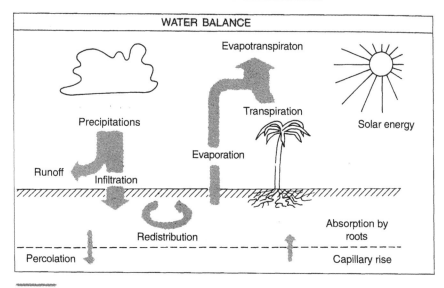

F(g 3.4   The principal terms of the water balance (after Musy and Soutter, 1991).

even more complex. This measurement is generally not available, either for plots (it actually assumes that one uses measurements that are difficult to obtain) or in the observational networks, or in meteorological data (this too is dependent on the plant cover to enable obtaining precise estimates on plots, based on regional estimates). One must therefore evaluate the actual evapotranspiration (AET) in relation to the only available data, the potential evapotranspiration (PET). For example, if the conditions of water supply resulting from the hydrological conditions of the soil are not limiting, and if the plant cover is homogeneous and dense, the system evapotranspires at AET and we write AET = PET. So, using several hypotheses, drainage at depth can be estimated. Further, analysis of equation (III) shows that the flow of nitrate depends on the flow of drained water, estimated earlier by the water balance, and average concentration of the soil solution during the particular period of time. Lastly, the flows calculated in this manner frequently reveal considerable spatial variation, that is, considerable quantitative differences from one point to another in an agricultural plot, for example with regard to points that are often proximate. An example of this spatial variability is given in Figure 3.5, which shows the drainage calculated by the water balance at 36 points of a truck-garden plot (salad) during three successive crops. The origins of this variability are often multiple (heterogeneity of irrigation on the soil, preferential flows etc.).

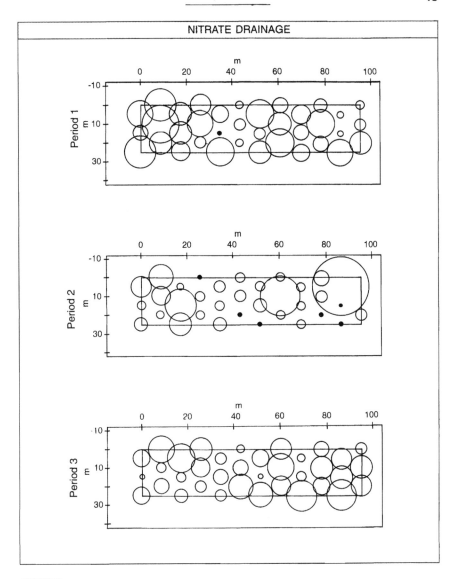

NITRATE DRAINAGE

Fig 35 Drainage of nitrate calculated by the water balance at 36 points on a plot in a vegetable garden (salad) in the course of three successive crops (period 1: spring crop; period 2: summer crop; period 3: autumn crop). The area of each cycle is proportional to the quantity of nitrate drained at the centre of the circle.

The water balance

Let us write the water balance of an agricultural plot.

The main entries are provided by the precipitations (P, mm) and irrigation (I, mm) while the losses correspond mainly to the actual evapotranspiration (AET, mm), drainage (D, mm) and the runoff (R, mm). If we use S_i and S_j to denote water reserve at times i and j (mm, with $i < j$), we immediately derive:

$$S_j = S_i + (P + I) - AET - D - R \qquad (\text{I})$$

Generally, expression (I) can be further simplified. For example, considering the topography of the plot, the runoff may be taken to be negligible. Then, from relationship (I) we can calculate drainage at any depth:

$$D = (S_j - S_i) + (P + I) - AET \qquad (\text{II})$$

Lastly, the quantity of nitrate drained may be estimated by the relationship:

$$D_n = D \times C \times 10, \qquad (\text{III})$$

in which C is the concentration of the soil solution ($N\_NO_3^-$ kg m^3), which is assumed to be constant between the times i and j, and the term 10 is the conversion factor for the different quantities (mm and m^3 ha^{-1}). The relationship (III) takes into account only convective flows associated directly with the carrier role of water that infiltrates the soil following a rainshower or excessive irrigation.

Estimation of flows in our environment at each time

The preceding method only suggests an overall balance of a situation between two time; the medium becomes poorer or is enriched, for example, but it provides no information on the dynamics of the events. Was there a particularly intense climatic incident (a storm) that explains the pollution of an aquifer, or is this a continuous process in the course of a year? What are the periods during the year when rain charges the water table or the water table provides water to the soil? Answers to such questions cannot always be provided by the water balance, but may be fundamental for explaining the essentials of the phenomena that the water balance describes only globally. To deal with these questions, an attempt is made to describe and quantify flows at each time based on the laws of physics (conduction, convection, diffusion) and on the properties of the soil (properties that describe the ability to store or transport matter or energy in the soil). These methods are more complex than the preceding ones because they demand a good knowledge of soil properties (as opposed to the method of balances, which only considers what enters and leaves the soil regardless of its basic properties), and also of the methods of computation, frequently important because the problems are often complex (recourse to data processing).

FROM THE INFINITELY SMALL TO THE INFINITELY LARGE

The various processes relative to the phenomena of transport in soils mentioned so far do not consider the time and space scales. The reason, it would seem, is that these processes exist simultaneously on a highly variable time (from a second to a century or a millennium) and variable space (from a square centimetre to a continental) scale. Fundamentally, the same types of processes are always involved, no matter the scales. What changes from one scale to another is the magnitude of the phenomena or the fact that an important phenomenon on one scale may become inconsequential on the other or vice versa. Thus, erosion of a soil layer (transport of solid particles) to a depth of a few millimetres may be negligible on this scale but become dominant on a geological time scale because it is this erosion that creates the reliefs. Increase of milligramme of NO_3 per litre in a water table may appear to be extremely small, but with such an increase, initially pure water will become unfit for human consumption in 50 years, if we apply the current norms for potability.

Time Scales

The phenomena of transport in soils are, as seen above, permanent and can be analysed from various points of view with respect to time scales. Let us take, for example, the case of heat transfer in soils. A small variation in air temperature is automatically transmitted to the soil surface in a few seconds. This temperature fluctuation is perceptible only in the first few centimetres of the soil because it is soon dampened. In the course of a day, the average change in soil temperature follows changes in the air temperature due to this phenomenon of damping (the temperature variations progressively decrease as we probe deeper) and the phase change (the time of maximum temperature in the soil at each depth differs from that of maximum temperature of the air). At the seasonal level, here again the temperature of the soil will reflect the temperature of the air and its variations. We may continue in this manner with much higher levels of time. Similarly, on the scale of a century or a millennium, any long-term variation in temperature will be reflected in the soil temperature, which will thus 'record' long-term climate changes. It can even be shown that there is a close relationship between the temporal and spatial scales: variations of temperature on the temporal scale of a second are recorded only over a few millimetres, variations in day temperatures over several tens of centimetres, seasonal variations over several metres and variations in centuries or millennia over depths of several hundreds of metres. This interesting property is also currently used as a method for determining past climates by recording temperatures at low depths (several hundreds of metres) and their variations. Thus an increase in temperature at a depth of a few tenths of a degree may reveal a warming of the air at the soil surface of several degrees a century earlier. The soil therefore operates as a veritable recording thermometer of past climates.

In the water cycle for example, processes also coexist on all temporal scales, but the utility of a particular process depends upon the actual time scales. Thus to estimate the drainage or runoff by water balance in an agricultural plot or a catchment area on the scale of a second, assuming it were possible to do so, would be meaningless. It is on an annual or decade scale that such a balance would be meaningful and significant with respect to management of the environment.

Spatial Scales

From a strictly local scale (at a point) to that of a continental or planetary one, the elementary processes are the same globally. Some sufficiently robust methods can also be used on different scales. This applies to the water balance which is based only on the law of conservation of mass (a universal law) and which is therefore applicable regardless of the volume studied. We may hence determine all the components of the water balance on a scale of a square metre, an agricultural plot, a small catchment area (a few hectares) or a very large one (for example, the catchment area of the Amazon) or even on a planetary scale. As and when the size of the studied area increases, we look for progressively global measures. For the Amazon basin, knowledge of evapotranspiration can only be regional and we may determine the amount of water lost from the catchment area by a global measurement of the flow of the river at the mouth where it meets the ocean. These global measurements often integrate extremely variable realities which change from one point to another. In fact, soil is sometimes fundamentally heterogeneous over very short distances. As a result, all observations relative to soil in general and to the phenomena of transport in soils in particular, are subject to this variability. Knowledge of a measurement can never provide a fair estimation of the average value of the phenomenon and we can only approach reality by recourse to statistics, which is always uncertain due to the likelihood of error. Thus rainfall may be extremely variable over several hundreds of metres and measurement of rainfall at any single point is not necessarily representative of an entire zone. On a cultivated plot, irrigation by the drip method may induce considerable heterogeneities in the supply of water from one point to another (from a simple increase or even more!).

Glossary

Conduction	Phenomenon of propagation from neighbouring points through a conductive element (for example, conduction of heat).
Convection	(From the latin *cum* meaning *with* and *vector* which means vector in English). Movement corresponding to the displacement of a particular element following the movement of an entire liquid or gaseous phase (for example, transport of nitrate in the liquid phase by convection).
Diffusion	Movement that occurs for a particular element in a liquid or gaseous phase when the phase does not have a uniform concentration at different points (for example, transport of nitrate by diffusion).
Dispersion	Phenomenon by which a provision is highly localised at a point in a particular element (water, solute, gas) and spreads eventually over a progressively wider space in the course of time.
Evapotranspiration	Total direct evaporation of water from the soil surface and transpiration through plants, that is, the water extracted from the soil by roots, which passes through a plant and evaporates into the atmosphere through the leaves. Potential evapotranspiration corresponds to the quantity of water that may be lost by the plant cover when it is well supplied with water. Actual evaporation is equal to the quantity of water that is lost by the plant cover which may be less than the potential evapotranspiration if the water supply is limited.
Nitrate	One of the ionic mineral forms in which nitrogen exists (it is essentially in this form that nitrogen is absorbed by plants).
Pollution of water table	All the phenomena of chemical pollution which affect the quality of water in aquifers, of agricultural or urban origin (pollution by nitrate, heavy metals, organic molecules from plant protection products used in agriculture).
Soil horizons	These are different layers of soil which are differentiated according to their origin and properties.

Soil porosity	Volume of soil filled with water or air.
Soil solution	All the water contained in soil and the elements dissolved in it.
Water balance	Balance of gains, losses and variations in the water reserve in a given volume of soil. The water balance is based on the law of conservation of mass.

FURTHER READING

Agassi M. 1996. Soil Erosion, Conservation and Rehabilitation. Marcel Dekker, Inc., NY–Basel–Hong Kong, 402 pp.

Don Kirkham, Powers WL. 1972. Advanced Soil Physics. Wiley Interscience, John Wiley & Sons, Inc., NY–London–Sydney–Toronto, 534 pp.

Hanks RJ. 1992. Applied Soil Physics, Soil Water and Temperature Applications. Springer-Verlag, NY–Berlin–Heidelberg–London–Paris–Tokyo–Hong Kong–Barcelona–Budapest, 2nd ed., 176 pp.

Hillel D. 1971. Soil and Water, Physical Principles and Processes. Academic Press, NY–London, 288 pp.

Jury WA, Garner WR, Gardner WH. 1991. Soil Physics. John Wiley & Sons, Inc., NY–Chichester–Brisbane–Toronto–Singapore, 5th ed., 328 pp.

Musy A, Soutter M. 1991. Physique du sol. Presses polytechniques et universitaires romandes. Collection gérer l'environnement, 335 pp.

Tardy Y. 1986. Le cycle de l'eau. Climats, paléoclimats et géochimie globale. Masson, Paris–NY–Barcelona–Milan–Mexico–Sao Paulo, 338 pp.

Yaron B, Calvet R, Prost R. 1996. Soil Pollution, Processes and Dynamics. Springer Verlag, NY–London, 310 pp.

CHAPTER 4

Soil and Exchanges with the Plant Cover

J.C. Fardeau, P. Stengel

Soil is described as the somewhat friable layer on the surface of the earth's crust. In climatically appropriate conditions, various plant species that form the plant cover grow on it and die. They remain in place in natural ecosystems or are exploited as food for humans or animals in agrosystems. The soil, constituted of a solid, mineral and organic phase with pores and all the living organisms continuously found in it, is the explored, exploited or exploitable zone used by the roots of the plant cover in the first metre, or even less, of the earth's crust. This is why by plant cover should be understood not only the aerial parts of the plants, but also the roots. The soil-plant cover couple can be considered a system with two components between which there is a permanent exchange of matter. In fact, the plant cover, which is inevitably temporary albeit living, develops at the expense of the soil that provides the nutritive elements, while soils develop partly by incorporation of a fraction of the organic matter produced by the plant cover. The soil, so described, is the interface between parent rock, eventually between the groundwater table and plants, the incontrovertible basis of our food, but the soil-plant cover couple may also be considered an interface between stone, the parent rock and the atmosphere, an equally indispensable component of the soil and the plant cover.

TECHNIQUES FOR QUANTIFICATION OF EXCHANGES AND TRANSFERS

For analysis of transfer of matter between soil and the plant cover two principal techniques are used. These are the method of balances and techniques based on the use of isotopic tracers.

Method of Balances

This consists of determining the variations in size of each component over a period of time. It is easy, for example, to measure the increase in quantities of an element within a crop obtained from a soil that was initially bare. But this method has its limits; it can hardly be used in such ecosystems as permanent grasslands in which the temporal variations are very limited. Generally, it is difficult, even impossible, to adopt this method whenever the soil and plant cover components of the studied systems are in dynamic equilibrium, that is, when there are exchanges between them of quantities of matter in the same amounts and in opposite directions or whenever the quantity of an element transferred from one component to another is limited compared to its mass in each of the components.

Isotopic Tracers

Isotopic tracers are the preferred tools whenever the system in which it is desired to analyse transfers is formed of several components and the transferred quantities are very limited compared to the quantities present in the receiver components. A component is defined as an assemblage in which the constituents have identical or similar properties, especially the ones related to transfer. The plant cover is a sufficiently organised component so that life-sustaining processes can occur even when the soil is much less organised compared to the living world. This is why soil and plant cover may be considered two different components which exchange matter between themselves.

In natural ecosystems, such as forests or permanent grasslands, the quantities of elements involved each year, whatever be the elements, are generally small compared to their quantities in the soil and the plant cover. An analysis of the total elements in the aerial part at two different periods cannot help us to know precisely the quantities that are extracted every year. This is why, in such situations, it is essential to make use of isotopic tracers which initially help to 'illustrate' the transfers and later to quantify them. Very many tracers have been used to analyse functioning of the soil-plant cover systems. Aside from the most common isotopes in the natural state, these tracers are: ^2H or ^3H for a follow-up of water, ^{13}C or ^{14}C for studying the carbon cycle and ^{15}N for the nitrogen cycle, ^{32}P or ^{33}P and ^{40}K to precisely determine the kinetic properties of phosphorus and potassium assimilated from soils. Some isotopes are stable, for example ^2H, ^{13}C or ^{15}N, others are radioactive, such as ^{32}P or ^{33}P.

The following diagram (Fig. 4.1) is an actual and classic example of isotopic tracking.

Let us use a phosphatic fertiliser P at the rate of 50 kg per hectare, 'labelled' with a certain quantity of radioactivity, R. At the time of harvest, considering the reduced radioactivity, we shall find, for example, that in a plant the radioactivity is R/10. The actual coefficient of utilisation of phosphorus in the fertiliser is 10%, which corresponds to 5 kg of phosphorus in the plants that was received from the fertiliser. Thus, of the 50 kg supplied, 45 kg have remained in the soil after the first crop and continue to react with the soil constituents. A chemical analysis of the plants will reveal, for example, that the harvest contains 40 kg of P: 5 kg from the fertiliser and 35 kg from the assimilable phosphorus in the soil reserve. All or part of the original 45 kg from the fertiliser that remains in

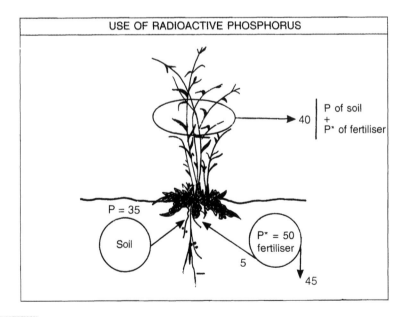

Fig. 4.1   Use of radioactive phosphorus to analyse the origin of phosphorus present in a plant.

the soil will replace the 35 kg of P extracted from the assimilable soil fraction. This very general outline illustrates that knowledge of only the 'entry-exit' balance of a soil can never help to understand the mechanism involved in soil-fertiliser-plant systems.

Compared to chemical determinations the use of radioactive isotope tracers such as ^{32}P, ^{33}P or ^{14}C helps to considerably increase the sensitivity in measurements of transfers. Detection of radioactive emission is proof of the presence of one actual atom, while chemical determinants correspond, in the most favourable cases, to the presence of about 10^{10} actual atoms. The precision of these measurements is limited due to the fact that radioactive energy emission is a random phenomenon. In practice, this is generally less than 3% and precision cannot be better than 1%.

The use of isotopes is based on the hypothesis that the fate of two isotopes of the same element is identical. This is verified whenever it is not possible to show differences of less than 1% in their kinetic behaviours, that is, for all pairs comprising one radioactive and one stable isotope or two radioactive isotopes. In fact, every transfer from one compartment to another or from one chemical form to another, is a kinetic process. Schematically, the heavier isotopes move somewhat slower than the lighter ones. Repetition of these processes during sequential biological reactions results in a change in the relationship of the two isotopes in the compartment of their origin and also in the receiver compartment: this is called the isotopic discrimination and variations in natural contents. This discrimination is always very limited. It can only be observed by the use of mass spectrometers and measuring not just the absolute quantities of the two isotopes, but the ratio between the two isotopes of the same element. This type of approach may be used whenever an element has at least two stable isotopes in the natural state. It particularly applies to carbon with ^{12}C and ^{13}C, nitrogen with ^{14}N and ^{15}N and oxygen with ^{16}O, ^{17}O and ^{18}O in which the natural variations are used to track the fate of organic matter, the origin of nitrate pollution in phreatic water and the water cycle.

EXCHANGES AND TRANSFERS BETWEEN SOIL AND PLANT COVER

Transfer Pathways

There are two transfer pathways between these compartments. They are, on the one hand, the roots and their auxiliaries, namely bacteria and fungi associated with the roots in most situations. Roots are successively conduits that enable soil → plant cover and plant cover → soil transfers and cellular structures in the soil. These are, on the other hand, all dead aerial parts of the plant cover which, like dead roots, return to the soil at the time of maturation of the cover and later release mineral elements or organic compounds.

Transfer Mechanisms: Nature and Function of the Transferred Constituents

Direction: plant cover → soil

These transfers are a result of growth, at the expense of atmospheric CO_2 and activity of the plant cover. Firstly, the roots, like any living organisms, breathe and release CO_2 in the soil atmosphere. The majority of carbon transfers to the soil are of a different origin. The cover 'injects' a part of the carbon compounds synthesised in its leaves into the soil. These are either carbon compounds, such as mucilages and mucigels excreted during growth of the cover from cells in the apical meristem of the roots, or compounds released by microbial decomposition after the death of the roots. The excreted compounds are essentially sugars. They soon serve as a substrate for bacteria of the rhizosphere which, in turn, release into the medium compounds such as polysaccharides responsible for aggregation of soil particles or exoenzymes that are likely to hydrolyse many organic compounds present in soils. After root mortality, the constituents are compounds from the degradation of cell walls, such as cellulose, lignin and other constituents contained in the cells, in particular potassium, phosphates and organic compounds of nitrogen which subsequently, after hydrolysis, release ammonium and CO_2.

Estimation of the amount of carbon compounds 'injected' into the soil is not possible even on an annual basis by measurements of quantities of carbon present in the soil before and after a crop. Two factors render this estimate erroneous. On the one hand, the quantity of organic carbon initially present in the soil is always considerable compared to that injected during the period when the root system is functional; on the other hand, a fraction of the organic carbon is consumed by soil microflora and released rapidly into the root system in the form of CO_2 and therefore is not directly measurable. Only the use of CO_2 labelled with radioactive ^{14}C or more recently with stable ^{13}C, in the atmosphere where photosynthesis occurs, enables precise measurement of transfers of carbon compounds to the soil, whether they be excreted compounds or their derivatives because of microbial metabolism. At the beginning of growth the quantity of carbon transferred from the plant cover, including the roots, is 30 to 35% of the photosynthesised quantity but at the harvest, or more generally at ripening, it is only 10 to 15% of the carbon fixed by photosynthesis.

The processes of mineral nutrition also cause a release in the soil of protons of two origins: on the one hand, the 'proton pump' operates to allow the entry of cations in the root; on the other hand, all nitrogenous nutrition based on nitrates in the course of transformation of nitrates to ammonium within the cover, induces the production of protons which are released in the soil from the roots.

Further, all plant cover, whether cultivated or natural, loses its leaves that inevitably fall and are then decomposed by all the living organisms in the soil. In this way, there is a return to the soil of not only the carbon substances of photosynthetic origin and previously extracted minerals, but also compounds or constituents trapped by the leaves during their growth. These trapped products are either particulate atmospheric deposits of fairly distant origin that bear no relationship to agriculture, such as radionucleids from Chernobyl in 1985 or clays. They may be constituents that eventually help development of the cover or substances in solution such as sulphates or protons of industrial origin, or even environmental pollutants, toxic or otherwise, deliberately introduced. This, for example, applies to pesticides; but in the case of pesticides it often happens that most of the amount added does not reach the target during its application and falls directly on the soil.

Direction: soil → plant cover

These transfers are the consequence and an illustration of root exploitation. They are, on the one hand, substances in transit in plants: water, and on the other hand, mineral elements necessary for synthesis of the plant cover.

—*Water*: To increase the dry matter content by 1 g a plant, depending on climatic and nutritional conditions, absorbs and then transpires between 300 ml and 1 litre of water. At the final phase it has retained not more than 0.5% of this water. The remaining 99.5% is released into the atmosphere. For its build-up therefore, a plant cover needs a large transfer of water from the soil solution to the atmosphere.

—*Mineral elements*: These are extracted by plants from the soil solution in an ionic form or, as an exception in chelated form, by very small organic molecules of several toxic or non-toxic trace elements. The extracted quantities at a particular moment are proportional to the concentration of the elements in the soil solution. Three mechanisms operate simultaneously in the transfer. These are mass flow or mass transport by the water of transpiration, diffusion and desorption from the solid phase to the liquid phase. The three mechanisms are explained below using the example of phosphorus extraction by plants and extended to other elements.

The concentration of phosphorus in the form of phosphate ions in a soil solution is, on average, in soils that have a long agricultural tradition as in Western Europe, 0.2 mg phosphorus per litre of soil solution. This concentration may vary from 0.002 mg per litre in tropical soils cultivated over a long period with no replenishment of phosphorus, to 2 mg phosphorus per litre in soils that receive or have received large amounts of phosphate fertilisers or slurry from piggeries. In standard soil and agricultural conditions of Western Europe an increase of 3 g dry matter will involve an extraction of about 1 litre of soil solution. This litre of soil solution will automatically carry in its water current, that is, **mass flow**, an average of 0.2 mg of phosphorus into the plant. Analysis of the 3 g of dry matter revealed that the plants had concomitantly extracted 9 mg or even 10 mg of phosphorus. The mass flow therefore accounts for only 0.2/9, that is, 2 to 3% of phosphorus extractions.

There is no exception to this result. This leads to the conclusion that the rate of entry of phosphate ions in the root is 40 to 50 times higher than the rate of penetration of water molecules and that a root may select the elements entering it: root absorption is an **active process**. The laws of physics indicate that such a transfer occurs only from a high to a lower concentration, that is to say, by **diffusion**. If we next compare the quantity of water present in a soil, its concentration and the quantity of phosphorus extracted, this result also leads to the conclusion that 90 to 93% of the extracted phosphorus must leave the solid phase of the soil and be desorbed during the growth phase to ensure phosphate nutrition. Extraction of phosphorus from the soil solution by the plant roots causes a transfer, a **desorption** or a **dissolution** of phosphate ions from the solid phase of the soil to the root through the soil solution. This result, largely valid for any soil, illustrates that soil behaves as a vast reservoir of biologically available phosphorus for plants. Knowledge of the mechanisms that govern the functioning of this reservoir is essential for controlling nutrition of the plant cover.

Similar measures and computations have been adopted for other elements extracted by the plant cover. The concentration of potassium in the soil solution is 10 to 15 times higher than that of phosphorus but the extraction is also significantly higher: mass transport in the transpiration flow accounts for 15 to 20% of the extraction. In the case of nitrogen this mechanism, depending on years, accounts for 20 to 50% of the nitrogen extractions. The case of calcium is also of interest. Actually, in calcareous soils its content in the plants, considering the concentration of calcium ions in the soil solution and the water flows, should be 150 to 200% of the observed values. Schematically, the contents of major nutritive elements in the cover are more than those in the medium from which the cover draws its nutrients: on the other hand, in the very same plant cover there may be a reduction in the contents of other elements. These observations confirm that a plant cover behaves as a selective filter which modifies the relative contents of elements in the soil and cover compartments.

Transferred Quantities and Kinetics of Transfer

Direction: plant cover → soil

The distribution of dry matter between the aerial and the underground parts of a plant cover depends on the type of plants. To be convinced of this, it suffices to consider the sugar beet or potato and wheat or maize.

The total quantity of carbon transferred from the plant cover via the roots and including the roots to the soil compartment depends mainly on the production of dry matter, in other words the efficiency of photosynthesis. It is, on average, 25% of the quantity of photosynthesised carbon.

Transfer of carbon to the soil in the form of dead aboveground plant parts is highly variable. It depends firstly on the nature of the plant cover and its productivity. The quantity of dry matter produced every year varies depending on the ecosystem or the agrosystem from 0.5 t per hectare to 25 t per hectare. Thus, for example, in Togo, on soils cultivated for 25 years or more which have never received fertilising elements and in which even the crop residues have never been ploughed back but are always taken away from the cultivated plots and used for non-agricultural purposes, the grain yield of maize now is never more than 0.5 t per hectare, in other words, a total production of dry matter of about 1 ton. At the same time, the cultivated maize yield in the southern part of France may reach 14 t ha^{-1} y^{-1}, i.e., 30 tons of dry matter. These transfers also depend on the modalities of managing the residues of the harvest because in Togo the entire dry matter produced, except for the roots, is removed from the plot for use as fuel or cattle feed, while in southern France 16 tons of dry matter are ploughed back every year.

The return to soils of mineral elements associated with dry matter depends on the quantity of dry matter and the mineral element content in the residues. Thus in the case of potassium they will be nil in Togo and as much as 150 kg in southern France.

Direction: soil → plant cover

The quantity of mineral elements transferred from the soil compartment to the plant cover compartment depends, on the one hand, on the quantity of dry matter produced and, on the other, on the content of the given element in the dry matter. The content of elements in the cover depends on the element, the soil and, in the case of a particular soil, on the quantity of nutritive elements provided to the crop, and the climatic conditions which may influence the transformation of non-assimilable organic nitrogen in the soil to biologically available mineral nitrogen. It depends also on the plant, its age and, within a plant, on the organ. Table 4.1 illustrates these observations and provides several ranges of contents for several elements which are necessary for the development of plant covers.

Table 4.1 Contents of some elements in a few plants. K, P, Ca are given in g kg^{-1} dry matter and Fe in mg kg^{-1} dry matter

Plant cover	Plant part	K	P	Ca	Fe
Wheat	Straw	2.6 – 15.4	0.3 – 1.7	0.8 - 4.3	60 – 630
	Grain	2.3 – 9.7	1.5 – 5.4	0.05 - 3.0	30 – 420
Maize	Straw	2.6 – 19.0	0.4 – 4.2	1.0 – 8.4	310 – 320
	Grain	2.2 – 9.2	2.3 – 8.0	0.06 – 0.6	13 – 550
Sugar beet	Leaf	17.0 – 65.0	0.9 - 4.4	8.0 – 31.0	142 – 1990
	Root	3.7 – 45.0	0.35 – 6.2	0.9 – 28.0	69 – 290
Soy bean	Seeds	8.0 – 24.0	5.0 – 18.0	1.2 - 3.4	50 – 160

Using this type of informative observations, agronomists have begun to forecast the requirements of nutritive elements for cultivated plant covers. The requirements of a cover are computed taking into account the content of the element and the expected crop yield. Further, based on the assumption that tools are available to estimate the quantity of the element which the soil can provide to the cover and that the fate of the fertiliser is known, it appears possible to calculate the quantity of the element to be provided in the form of fertiliser to obtain the expected yield.

The differences between estimation of the requirements and effective extraction are very often positive. Many factors may explain these differences. Some are objective: the content of elements retained for computations are often the highest observed values. Others are more subjective and are associated with the attitude of farmers or their advisers: on the one hand, the expected yields are estimated without considering climatic uncertainties, especially the precipitation; on the other hand, a farmer does not want to take a risk and adopts an attitude of assurance based on the principle that, in a situation where there is a small excess, none of the elements provided will be toxic for the cover. In other words, loss of yield will not result.

The time lag between estimation and extraction also raises two questions: Are the tools available for forecasting soil-plant cover transfer effective? Do we know what happens to the fertilisers? These two aspects are dealt with later.

Nevertheless, this step, howsoever imperfect, indicates genuine progress compared to earlier methods based essentially on empirical data obtained during in-field determination of functions associated with the yields of the cover or even the extrapolation of these data to agropedological situations different from those where they had been obtained. The most caricatural case in France is certainly that of determination of the nitrogen requirement for wheat. Based on field trials, the formula '30 kg of nitrogen per ton of wheat' was popularised. The actual consumption by wheat, in other words its effective requirement, is generally never more than 20 kg nitrogen per ton. Such a simplified formula, which is not based on the mechanisms that operate in the soil and plant cover, has markedly contributed to loading the phreatic water with nitric nitrogen; leaching of nitrate to phreatic waters was in fact implicitly included in the estimated 30 kg N requirement.

The total quantity of an element that a cover may extract is not the only information required. Not only does this quantity vary depending on many parameters but also for a given quantity to be extracted to ensure production of the expected dry matter, the kinetics of extraction varies depending on the plants. The duration of extraction is generally shorter than the period of growth and varies with the elements (Fig. 4.2).

Figure 4.2 provides much information on transfers and exchanges between the crop cover and soil and the difficulty in interpreting analytical data that would obtain from a single sampling. Thus, for wheat:

- the quantity of dry matter in the aboveground part decreases at the end of growth. This actually corresponds to the dropping of old and senescent leaves on the soil.
- net transfer of potassium in the soil→plant cover direction ends when anthesis begins. Subsequently, net transfers in the plant cover→soil direction begin. Later on redistributions among the different parts take place.
- very large quantity of this element returns to the soil during the maturation stage of wheat.
- transfer of phosphorus and nitrogen in the soil→plant cover direction continues long after that of potassium.

Thus, it is clear that it may often be difficult, even impossible, to superpose the kinetics of release of nutritive elements from the soil on that of extraction by the plant cover. Whenever the objective is to obtain the maximum crop yield, this situation almost inevitably leads to

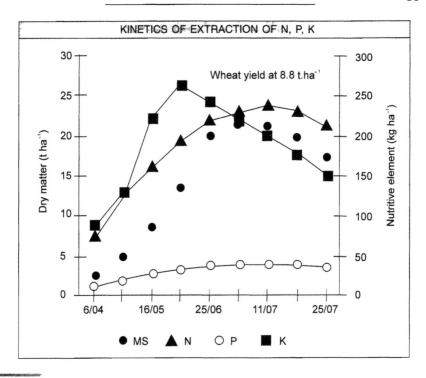

KINETICS OF EXTRACTION OF N, P, K

Wheat yield at 8.8 t.ha^{-1}

Dry matter (t ha^{-1})

Nutritive element (kg ha^{-1})

● MS ▲ N ○ P ■ K

Fig. 4.2 Kinetics of extraction of phosphorus, potassium and nitrogen for wheat and accumulation of dry matter (DM) (Data SCPA—Aspach).

provision of more fertiliser than actually necessary to meet the requirements of the plant cover. These observations *ipso facto* justify the view that some fertilisers, especially nitrogen fertilisers, which may be leached by rain-water, should be provided in installments when the plants most need them.

The existence of different extraction kinetics for the same element helps explain in part the empirical division of cultivated plants into three categories: Those with a large requirement, those with a moderate requirement and those with a limited requirement for one or several elements (Fig. 4.3).

The mechanisms that would help to fully understand the concept of requirement for covers are not yet known, especially since some plants are capable of extracting certain elements from the soil in quantities that are much higher than those actually necessary to obtain the desired yield. This applies to lucerne, for example, with respect to extraction of potassium. The consumption is considered extremely high and to date no explanation has been found for it.

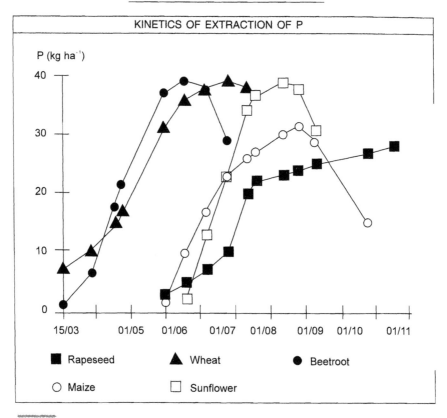

Fig. 9.) Kinetics of extraction of phosphorus by 5 different plants (Data SCPA—Aspach).

Cycle of Elements between Soil and Plant Cover: Biogeochemical Cycle

In a natural medium

The plant cover extracts elements from the soil in a biologically available ionic form but releases them to the soil in the form of structures of plant origin containing nutritive elements included structurally in organic compounds, or in the form of organic compounds with root exudates. This situation is especially true with respect to nitrogen, the basis of production of amino acids and proteins, and for phosphorus, the basis for the constitution of DNA and RNA, for example. These structures and compounds are not in a biologically available state for plant covers which may be cultivated later on. The cycle of the elements can therefore be completed only with the intervention of other living organisms:

herbivorous vertebrates, herbivorous invertebrates and detritovorous decomposers participate, more or less equitably in the plant decomposition, depending on the ecosystem. Next, soil micro-organisms transform part of the organic compounds into biologically available elements. It has been reported that the concentration of nutritive elements is generally higher in plants than in the soil where they are grown. During transformations, due to the enzymatic activity of micro-organisms, the released elements are not remixed uniformly in the soil: they remain more or less localised at the point of contact of the plant structure. Repetition of such a cycle over millennia has contributed to an increase in the concentration of available elements at the surface and in the layer explored by roots.

In a cultivated medium

In an agricultural field the situation is slightly different. Actually, agriculture produces food which is transported to towns along with the elements contained in it. At the level of the cultivated field the cycle becomes open. To complete it, at least quantitatively, the farmer must, at the very least, restore the quantity of elements removed. There are two possibilities: recycling waste resulting from human consumption or providing fresh nutritive elements, called fertilisers, extracted from other sources or directly from mines for P, K or Ca and even from energy-consuming factories for N. The source for waste hardly differs in concept from that encountered in the natural medium. Nevertheless, during production and subsequent collection of residues from human consumption, xenobiotic products that are generally formed during industrial production are often found in human waste. Whether these are wastes or fertilisers, it is necessary to know if the restored form is biologically available for the plant cover at the time of its provision. The answer is sometimes yes, sometimes no. Whichever the case, the restored form reacts with the constituents and living organisms in the soil and is transformed.

CONCEPT OF SOIL FERTILITY

This has long been judged by the appearance of plants that grow in the soil. A plant cover that is pleasant to the eye may be capable of accounting for easy exchange and transfer between a fertile soil and its cover. Can a visual assessment, however, suffice to quantify fertility of a soil? Can it replace and eliminate a more detailed analysis of the soil, plant cover and the conditions of functioning in this association? Naturally the answer is no, simply because it is always a very risky proposition to base a judgement on an object, in this case the soil, only through another object, in this case, the cover: the prism may be a highly deforming one.

The concept of soil fertility may be analysed fairly objectively by examining, for example, the increase in wheat yields during the last 100 years in France. These yields were stable at approximately 2 tonnes per hectare, on average, until about 1920. After that they reached 65 quintals per hectare on the same soil. Actually, the soil cannot provide consumable items continuously and in sufficient quantity without reasonable and continuously renewed human intervention. Thus, what would happen to grains of wheat simply broadcast and not buried in a soil prepared earlier to receive them? This indicates that fertility cannot be expressed or concretised without human intervention through human, animal or fossil energy, depending on the place. The increase in wheat yields in France illustrate in particular that, among the factors affecting yields, the aspect of native fertility of agricultural soil varies in a particular region depending on its level of economic development. Comparison of yields from the same plant between various countries also demonstrates that this aspect of native fertility varies from country to country.

Finally, fertility of cultivated soil is a notion inherited from the time when work was the only available factor in agricultural production and consequently obtained a productivity that depended closely upon the limiting factors imposed by the physical medium.

FORECASTING POSSIBILITIES FOR TRANSFER FROM SOIL TO PLANT COVER

Interest in this forecast is constantly increasing because it helps adjust the application of fertilisers according to plant requirements. The elements extracted by the plant cover are, by definition, in their biologically available form. The plant cover can extract them only in this form. The total quantity of elements present in the soil horizon inhabited and exploited by roots is generally much more than necessary to meet plant requirements. Nevertheless, shortages of many nutritive elements may limit the production of dry matter and hence hamper the best use of the solar energy received by the cover in the photosynthetic process. Whatever the element, be it nutritive or toxic, and whatever the plant, there is never a relationship between the total quantity of an element in the soil and the quantity that may be extracted by the cover. The only exception to this relationship is that if an element is not present in a soil, the plants will not contain it. An element cannot be 'synthesised' by plants from other elements: plants do not carry out nuclear transformations. The absence of a relationship between the total content of the element and the extracted element leads to an ambiguous conclusion that the total quantity of an element does not ensure a uniform availability for the cover. The biologically and physiologically available fraction to plants is that which can be transferred to the soil solution from where it can later be taken in by the roots of the plant cover.

To forecast the possibilities of transfer of elements, be they nutritive or toxic, from the soil to the plant cover, tools are required to estimate the biologically available elements present in a soil and their biological availability. The exchange and desorption of ions, the dissolution of mineral constituents, the degradation of organic compounds in the case of nitrogen and sulphur, known generally as mineralisation, are the mechanisms to be considered in explaining the nutrition of plant covers. The methods adopted to analyse soils must simulate one, or even two of their actions.

Cations

The principal cations involved in the build-up of dry matter in a plant cover, namely K^+, Ca^{2+}, Mg^{2+}, Na^+, are most often extracted from the soil by the mechanism of ion exchange. The operation involves stirring some soil in the presence of salts containing ammonium NH_4^+ and then determining the number of ions that have left the solid phase of the soil to return to the ammoniacal solution. As for ammonium, it is extracted by a solution of potassium salt. There are variants of the technique, depending on the region, but the principle is identical worldwide.

Anions

No single method for anions is recognised by the International Community of Agronomists. This is because of the nature of the mechanisms that help to explain the release of nitrate, sulphate, and phosphate ions from the solid phase of the soil. What happens to the first two of these elements basically depends upon the microbial life in the soil. So little is known about this that so far it has not been possible to think of a method simulating the sequential reactions involved. This is why routine analyses are limited to extraction from water of nitrates for nitrogen and sulphate ions for sulphur although it is well known that this quantity represents only 10 to 50% of that extracted during a growth cycle. For phosphorus, that is, phosphate ions, there are, on average, two methods of extraction in every country, not one of which is unanimously accepted in any country. The main reason for this situation is that the extractants used alter the relationships between phosphate ions and the solid phosphate phase during extraction.

Conceptual Limits of Extraction Methods

Mineral nutrition of the plant cover is both a selective and kinetic process. Even though the routinely adopted methods extract elements that are actually assimilable by plants, these are only extractions made during a fixed period. They cannot simulate and explain either the kinetic process characteristic of plant nutrition, nor the selective process, nor in the case of a given element the existence of biologically available variants. This is why, conscious of these limitations, researchers have developed more informative methods for potassium and phosphorus. It is possible to use electro-ultrafiltration for potassium: it extracts potassium ions from the soil, depending on their energy of binding, using variable electric currents yet tells us nothing about the kinetics of release. On the other hand, for phosphorus, the isotope technique for estimation enables precise determination of the kinetics of potential transfers of phosphate ions from the solid phase to the liquid phase in all types of soils, cultivated or otherwise. This is the method of kinetics of isotopic exchange of phosphate ions between liquid and solid phases.

Lastly, it should be kept in mind that there will be no transfer to a plant cover unless there are functional roots. This why the methods of soil analysis cannot be really useful unless combined with a knowledge of the root systems in the cover and their functioning. Based on this we can assess the possibilities for improvement in our knowledge of soil-plant cover relationships.

MODIFYING TRANSFERS FROM SOIL TO PLANT COVER: CULTURAL PRACTICES AND 'SUSTAINABLE' AGRICULTURE

All transfers from soil to plant cover require two successive mechanisms. The elements must first be released from the solid phase of the soil to enter the liquid phase and then be extracted by the plant cover. Changes in soil-plant cover transfers may then be obtained by acting on either of these mechanisms. Every modification in transfer from the soil to the plant cover produces a change of transfer in the opposite direction, that is, the entry of organic matter into the soil.

Controlling Transfer by Modifying the Ability of Soil to Release Biologically Available Elements

Effect of extraction without restoration

All agricultural production undertaken to provide food for man or animals will result in a transport of nutritive elements from the production plot to the source of consumption, generally a town. Such an operation, repeated for many many years reduces the possibilities for transfer of nutritive elements from the soil to the plant cover. This practice is adopted unintentionally in numerous developing countries which are presently, for economic reasons, not in a position to make use of fertilisers. The limitation of transfer from soil to plant cover is observed after a time, already reached in many developing countries by a drastic restriction in yields. With respect to food production, this fact reminds us that the soil should never be considered an inexhaustible repository to be exploited continuously and unabashedly.

Reduce growth limiting shortages

A shortage of nutritive elements, except in the case of total absence, cannot be defined in absolute terms: It is only a shortage based on the requirement of the plant cover. This requirement must therefore be estimated to establish the extent of the shortage. This shortage will be rectified by a reasonable supply of the required element and corresponds in agricultural practice to irrigation and the use of fertilisers.

Drought, at a global level, is the main factor limiting plant production. Water intervenes in two ways: on the one hand, synthesis of the plant biomass is always associated with a large amount of evapotranspiration compared to the quantity of dry matter produced; on the other hand, mineral nutrition through ions can only take place in the presence of water. Irrigation helps to regulate the water requirements.

In 1994, French agriculture, to mention only the major elements, utilised about 2 million tonnes of nitrogen, 0.6 million tonnes of phosphorus and 1.3 million tonnes of potassium, representing approximately 10 million tonnes of fertiliser or, on average, 300 kg of matter per hectare of arable agricultural area. The entire quantity of elements provided to the soil is not transferred to the plant cover. Thus, during the year following spreading, the average transfer of nitrogen is about 50% of the supply, of phosphorus not more than 12 to 15% and that of potassium about 30%. Hence, irrespective of the element, most of the supplied quantity does not reach its target, which is the plant cover. These results can be explained as follows: (1) the root system of the plant cover is in direct contact with no more than 3 to 4% of the soil volume; (2) among all these elements there is biological competition for nitrogen and sulphur, and physicochemical competition between the soil and plant cover for phosphorus and potassium. This competition illustrates the existence of the buffer capacity in the soil. It is thus clear that the management of fertilising elements is not an easy matter and that ignorance of the soil's buffer capacity may lead to wrong estimates of the required provisions. This led agronomists to suggest management of supplies of nutritive elements based not on a knowledge of the functioning of the source of the element, which is the soil, but on the nutritional state of the plants. This knowledge was initially obtained through a foliar diagnosis, by analysing the total contents of the elements in the leaves. Without exception, these total contents are neither reliable indicators in the leaves nor in the soil. Thereafter attention was paid to early indicators of the nutritional state of major crops, especially for nitrogen, because it was considered desirable that farmers should avoid 'assurance' fertilisations. Such an indicator of early diagnosis, developed in France, termed Jubil ® is the nitrate content in juice extracted from the base of a wheat stem.

Apart from the traditionally controlled elements—water, nitrogen, phosphorus and potassium—there are others more difficult to control given the present state of techniques applicable at the level of agricultural practice. Among these is oxygen, indispensable for root respiration in the vast majority of plants. Root choking can only be effectively prevented by drainage.

Reduce excesses of biologically available elements

This objective actually covers two very different situations and necessitates a distinction between biologically available nutritive elements and biologically available toxic elements.

In one of these situations, nutritive elements such as nitrogen, phosphorus or potassium, which are non-toxic even in very large quantities, are present in far greater quantities than the cover can extract. They are thus likely to move to another compartment, generally the surface waters or those at depth and may contribute to their pollution or initiate processes of eutrophication. Reduction of these excess elements is ensured by a temporary discontinuation of their provision together with their extraction by crops and removal of the harvest from the plot. Due to the buffering power of the soil there are efficient but slow means for limiting fluxes and avoiding excess of nutritive elements in natural waters.

In another situation elements that are toxic for plants or even man and animals, whatever their effect on the plant cover, and at least partly in a state available for plants, are transferred to the cover and contaminate it. Three conceptually very different steps may be taken to eliminate or merely reduce excesses of these elements or biologically available compounds. The mechanical way is to remove the contaminated layer of earth; the chemical way involves modifying the availability of the pollutant by an additional non-polluting constituent. Lastly, the biological method, now called bioremediation, involves absorbing the contaminating elements by micro-organisms or plants in order to reduce their reserve or their biological availability. Only this last method, using the plant cover, is dealt with here. Some plants, called hyperaccumulators, have the ability to store certain heavy metals within themselves after chemically inactivating them. Researchers are now trying to make use of this property to eliminate pollution from soils contaminated by normally phytotoxic heavy metals.

Control of soil pH

Development in certain plants depends on the soil pH that controls the ionic ambience in the soil solution. Some are called calcicoles and others, calcifuges. In a carbonate medium, the pH of about 8.0 is due to the presence of carbonates of calcium and magnesium and cannot be altered by cultural practices. When there are no carbonates, the soil pH is neutral or acidic. The growth of plants is not affected by the pH itself but, as soon as the pH falls below 5.8, aluminium ions, toxic for many plants, may appear in the soil solution. It is always possible to raise the pH of a soil if carbonates are available, that is, limestone. The only problem to be resolved is the distance to be covered between the limestone quarry and the zone where the pH is to be modified, since zones that are clearly acid are the ones where there is no limestone. Easy to resolve in a developed environment because of the availability of mass transport, there is presently no solution for this problem in many countries of West Africa where the soils are highly acid and the acidity is greatly increased by urea, used as nitrogen fertiliser.

Modification of Plant Cover and Its Functioning

Another way of modifying the transfer is to modify the receiver or its functioning. This actually is a very ancient practice and corresponds to the adoption of crop rotations in a particular plot. Although not properly explained, this practice had, among others, the advantage of exploiting the soil by different root systems and by plants that do not extract nutritive elements in the same proportions. The best known case is that of the sugar beet-wheat rotation. The first plant has a taproot capable of exploiting the soil almost uniformly to a depth of 30-50 cm. The second plant has a creeping root that tends to extract most of its nutritive elements in the first 20 cm, although it is always possible to observe several roots of wheat at a depth of 1 metre. Further, the relationship between extractions of N/P/K is 100/16/260 for sugar beet and 100/18/120 for wheat.

Leaving fallow areas was a regular practice. The period of latency thus created between two crops generally enabled leguminous plants to fix a small amount of nitrogen. On the other hand, this practice also enabled the ploughing back of plant material in which the contents of nutritive elements were generally higher than that in the soil. Nevertheless, keeping land fallow is in no way a substitute for the provision of phosphate and potassium fertilisers to compensate for their removal at harvest time.

At the global level the main constraint is drought and the second is nitrogen. Drought must always be guarded against. Irrigation is an effective means that modifies the kinetics of transfer of nutritive elements between the compartments, including that of nitrogen to phreatic water!!! The harm caused by nitrogen will always be formidable.

Among the factors likely to alter the functioning of plant cover on a fairly long-term basis, it is important to mention the changes in CO_2 concentration in the atmosphere. This gas is often referred to as the gas with a greenhouse effect, which is likely to contribute to global warming, but CO_2 must first be considered the most important nutritive element for plants. That is why an increase in its content is indicated by an increase in the quantities of photosynthesised carbon. This increase, however, does not appear to result in an additional extraction of nitrogen even when the quantity of nitrogen is not a growth-limiting factor. The direct consequence of this is an increase in the C/N ratio is plants. It is presently very difficult to predict for plants the long-term consequences of undesirable increase in the CO_2 content in the terrestrial atmosphere.

'Sustainable' Agriculture

Agriculture on soil must continue to produce the essential elements of food indispensable for human existence. Control of exchange between the soil and plant cover will remain a matter of priority. Consideration of this imperative has contributed to the emergence of the concept of 'sustainable' agriculture. Despite this concept which appeared at the beginning of the 1980s, the definitions proposed for describing 'sustainable' agricultural practices are still highly variable. They often reflect the ideology of pressure groups that are enthusiastic about the notion. All of them have two points in common however:

- that agriculture should be able to meet the present food requirements of the world population;
- that the present cultural practices are not likely to limit the production of food requirements for future generations.

Such an approach places man at the centre of the ecological concept and implies that damage to the environment may result from present practices and that the soil is really a fragile interface that must be respected to ensure a future for soil-plant cover transfers.

For example, it could be 'an agriculture that must evolve indefinitely towards greater utility for mankind, increased effectiveness of natural resources, and a balance interfacing the environment which is favourable to man and other species'.

Schematically, this agriculture includes, to some extent, the old idea about land management as a 'good head of a family', but is extended to management of the entire planet. Considering that man is an integral part of the fundamental ecological process, there could be a change from 'sustainable agriculture' to 'ecologically viable agriculture'.

FRAGILITY OF SOIL INTERFACE WITH RESPECT TO ITS FUNCTION AS A HOME FOR, AND DEVELOPMENT OF PLANT COVER

Soil Interface Indispensable

Plants can certainly be grown in an inert medium, such as sand, by providing them the mineral elements necessary for growth. Nevertheless, such a practice would perhaps never help to produce a quantity of food sufficient to meet the increasing nutritional requirements of mankind. For this type of production, the major constraint is the requirement of energy necessary to obtain nutritive elements without xenobiotic elements and ensuring a continuous distribution, which is indispensable in the absence of any buffering power for the root medium. Further, under present technological conditions, the amount of energy required for production is more than the energy collected by the photosynthetic mechanisms: the energy balance is therefore negative and non-viable, and eventually not 'sustainable'. Soil has therefore to continue for several more centuries as the essential route for ensuring our food production.

Soil Interface Modified by Plant Cover and therefore by Cultural Practices

All agricultural practices affect the physical, chemical and biological functioning of the soil interface vis-a-vis the plant cover. This is even the objective of agricultural practices. They must either promote root exploration by tillage or improve the chemical functioning of the interface by providing nutritive elements, the chief ones being nitrogen, phosphorus, calcium, even magnesium, and, of course, water. There are very few practices that help control microbiological activity. Inoculation of legume seeds with nitrogen-fixing symbiotic bacteria, and perennial crops with mycorrhiza may be started however as significant examples.

Do Some Anthropogenic Practices Cause Irreversible Modification Incompatible with Long-term Maintenance of a Plant Cover that can be Consumed by Man?

In an ideal situation the soil should contain an adequate amount of nutritive elements in a form that can be transferred to the plant cover free of any xenobiotics likely to be taken up by the cover.

Erosion

Erosion is a 'natural' phenomenon. Erosion generally reduces the thickness of the soil exploited by the roots, and the total quantity of potentially available nutritive elements. It is a mechanism that modifies the relationship between fine and coarse particles, the finer particles generally being the first to be eroded. The consequence is modification in the relationship of nutritive elements and modification in the buffering power of the soil vis-a-vis the elements P and K. There is no long-term agricultural practice without a minimum of erosion. The practices which encourage it are known and, conversely, those which help to avoid it. The latter must therefore be adopted whenever compatible with the expected plant productions. The transformation of bedrock into soil takes several millennia. This is why, at the human level, such aggression against soils should be considered irreversible.

Changes in level of soil reserve

The soil is a reservoir of nutritive elements. Like any reservoir it should not be overfull to avoid a spillover from the soil interface, nor almost empty to avoid a 'breakdown' limiting development of the plant cover. Nevertheless, practices that cause an overfilling of the reservoir and those that drain it need not be considered irreversible with regard to problems of the soil as a place of exchange with the cover. It is also here that significant progress has recently been made in our knowledge of the functioning of plant-cover systems, which has important consequences for the strategies of fertiliser management.

Changes in reservoir functioning

The major and essential characteristic of this reservoir, with respect to its functioning in maintenance of the plant cover, is certainly its buffering power vis-a-vis the majority of elements necessary for the development of plants. This buffering power helps the cover survive even when external conditions become unfavourable. Everything must therefore be done to maintain and even in some cases, to improve the functioning of the buffering power. The constituents involved in this function vary according to the element.

For nitrogen and sulphur, only the organic matter compartment in the soil ensures this function on the express condition that the soil micro-organisms are able to mineralise and synthesise organic matter. Many agricultural practices, especially those that break the soil to encourage germination of seeds, assist in the processes of mineralisation compared to organisational ones: the medium and long-term consequence is a reduction in total content of organic matter and the nutritive power of the soil in respect of nitrogen. Ploughing back aboveground plant residues can only limit reduction; it does not inhibit it. Under the conditions of present agricultural practices in Europe there is only one general way to avoid reduction in organic reserve: this is the creation, for a period of 10 or more years, of permanent grasslands that help to rebuild an affected organic matter reserve. This latter observation, together with the fact that ploughing back aboveground plant residues does not appear to increase the content of organic matter, leads to the belief that only carbon compounds from the roots or those excreted at root level and subsequently transferred by bacterial populations, help increase organic matter content. Thus, in any system of rotation, excluding permanent grasslands, that is, in any agricultural production where animal husbandry is not included, reduction in the organic matter content of soils should be considered a certainty. But it is presently difficult to forecast equilibrium values for organic matter content since our knowledge of the long-term consequences of various agricultural practices is inadequate.

For potassium, only mineralogical clays can ensure the function of buffering power. This compartment can be changed only by removal of clays by the effect of erosion: such development will obviously be very slow. For phosphorus, the situation is intermediate between that of nitrogen and potassium: the buffering power is generally controlled by poorly crystallised minerals such as iron and aluminium oxides which are always present in the soil and by organic constituents associated with it, the nature of which is still not clear. In a given soil reduction in organic matter content is related with an increase in phosphorus buffering and fixing powers: however, the addition of coarse organic plant matter does not engender a reduction in this buffering power. This demonstrates that the kind of organic matter that intervenes are the compounds transformed by the soil microflora.

Adverse effect on quality of reservoir content

The purification (cleansing) power of the soil has very often been highly praised and highlighted to justify certain practices adopted for the elimination of wastes from our society. This excessive praise is certainly an error on the part of the media. Micro-organisms in the soil are certainly capable, under specific conditions, of denitrifying nitrates, that is, to transform them into non-polluting gaseous nitrogen. Micro-organisms are also able to partly degrade plant sanitation products such as herbicides 2-4D or atrazine. Nevertheless, some products of degradation are adsorbed on fine particles of earth and thus retained due to the buffering power of the soil; they may also remain in this state before being remobilised by micro-organisms and returned to the living cycle.

It is also because of the buffering power of the soil that the biological availability of xenobiotic metal pollutants such as cadmium, chromium and copper that may be introduced accidentally in a soil, is reduced. These elements are, however, only adsorbed. Their fate is regulated by physicochemical laws that control the processes of adsorption-desorption: the higher the increase in the charge of an adsorbent in an element, the lesser the energy required for its desorption. Hence it is advisable not to consider the soil a bottomless dustbin capable of swallowing and digesting all our wastes. If filled beyond its capacity the dustbin will 'overflow': toxic elements will move to the plant cover and even into our food. This risk is even more difficult to detect as some xenobiotics are non-toxic for plants but toxic for animals. In such situations modifications undergone by the soil may be considered as provisionally irreversible since extraction by the biological method is slow compared to the speed with which the soil can be charged.

Acidification of soils

This case is intermediate between the change in functioning and change in the content of nutritive reserve. Development in plant cover depends upon the soil pH. Many current agricultural practices cause acidification of the soil. Among these, mention must be made of ammonium fertilisers, especially of urea, increasingly being used throughout the world. Soil bacteria, the *Nitrosomas* and *Nictrobacters* transform these ammoniacal forms to nitrates by releasing protons in the soil which may take the place of calcium and magnesium on the absorption complex. These cations are then leached below the zone of action of the roots during periods when the availability of water, particularly from rainfall, is more than the plants can use, especially in winter. A reduction in pH occurs. Below a certain pH of about 5.0, biologically available aluminium, toxic for plants, appears in the solution by the breakdown of aluminium compounds in the soil. We have here a situation wherein the agronomist knows the causes of transformations and the possible remedies. Nevertheless a kilogram of ammoniacal nitrogen is cheaper to buy than a kilogram of nitrate nitrogen:

the agronomist is expected to resolve a problem not of his creation, but rather the result of remunerative industrial choices. This is generally not an irreversible transformation because an addition of calcium in the form of lime will usually help to correct the pH and remove the exchangeable aluminium.

Soil acidification is a very frequently observed phenomenon in a tropical climate even under natural conditions. It is rarely observed nowadays in cultivated fields of the temperate zone, but the appearance of acid rains, associated with industrialisation, has increased its probability. Difficult to estimate in an agricultural environment because of intensive human soil management, this process of soil acidification has contributed to long-term decrease of forest soil fertility in several European countries.

CONCLUSION

Plant cover is the only source, eventually through the animal, of our food. Its development results from transfer of nutritive elements extracted from the soil. This makes soil indispensable for many reasons.

Firstly, the soil is the physical substrate that ensures the planting and anchoring of crops followed by transfer to the plant cover. Formation of a soil as a result of pedological processes is an adventure that extends over several millennia. This is why, on a human scale, our soils, our lands, should never be considered renewable resources: erosion must be reduced to the minimum.

Next, the soil must remain friable to help penetration of the roots which are the essential pathways for transfer from the soil to the plant cover. This state depends mainly on the soil structure: that is, the organisation of clay particles among themselves, in the form of aggregates, thanks to original organic compounds formed directly or indirectly from photosynthetic processes. Soil tillage, the content and nature of organic compounds and the type of plant cover which will influence the return of organic compounds which influence to the soil, are the factors soil structure.

The soil is also the principal source of nutritive elements necessary for plant covers. All the elements are kept in reserve in it together with solid particles of the soil due to their properties of adsorption and buffering power. The functioning of this reserve depends upon the element. It is controlled entirely by microbiological processes in the case of nitrogen, by physicochemical process for potassium and by both for phosphates.

Lastly, the purificative power of the soil has frequently been highlighted to suggest that the soil is a repository for many of our potentially pollutant wastes. Nevertheless, for most of the mineral elements, use of the soil cleansing power is nothing but the effect of the soil's buffering power. The fixing power 'renders' these elements 'ineffective' for some time but does not actually eliminate them except in rare cases. The soil reservoir should not receive products alien to life, such as xenobiotics. The cleansing ability of the soil is a myth that must be rejected.

Glossary

Biogeochemical cycle	Cycle of an element in a natural ecosystem or agrosystem. This cycle illustrates the transfers and chemical transformations between the mineral or organic part of the soil and the living world.
Biological availability	Ability of an element to be transferred from a particular component of the soil to a living organism and be metabolised in it.
Durable agriculture	All agricultural practices that help meet the present nutritional requirements of humanity without an irreversible adverse effect on the production potential of soils. Such an agriculture is often called 'sustainable'. It is, in fact, 'ecologically viable' agriculture.
Fertility	Ability of a soil to provide expected yields.
Hyperaccumulators	Said of a living organism (plant, animal or microbe) capable of accumulating toxic elements within itself after having modified their chemical properties by a change in degree of oxidation or through chelation.
Mass flow	Transport of mineral elements into the roots through the water current from a plant cover.
Plant cover	All aboveground parts and roots developing on or in a soil.
Rhizosphere	A zone of soil subjected to the effect of plant roots. This zone contains mineral and organic elements of the soil that are likely to enter the roots, constituents excreted by the roots, including protons and such organic compounds as polysaccharides, micro-organisms associated with the roots as well as compounds resulting from their metabolism.
Yield	Quantity of matter produced per unit of area. In France the most frequently used unit of yield in agriculture is quintal per hectare or 100 kg per 10,000 m^2.

FURTHER READING

Decroux J, Ignazi JC. 1993. Matières organiques et agricultures. Comifer, Blois, France.

Duvigneaud P. 1974. La synthèse écologique. Doin, Paris.

Edwards CA et al. 1990. Sustainable Agricultural Systems. Soil and Water Conservation Society, Ankeny, Iowa, USA.

Jaillard B, Hinsinger P. 1993. Alimentation minérale des végétaux dans les sols. Encyclopédie des Techniques Agricoles, 1210: 1-13.

Loué A. 1987. Les oligo-éléments en agriculture. Agri-Nathan International, Paris.

Mengel K, Kirkby EA. 1987. Principles of Plant Nutrition. IPI, Basel.

Morel R. 1989. Les sols cultivés. Technique et documentation. Lavoisier, Paris.

Pontificiae Academia Scientiarum. 1968. Semaine d'étude sur le thème "Matière organique et fertilité des sols'. Wiley Interscience, NY.

Scharpensel HW, Schomaker M, Ayoub A. 1990. Soils on a Warmer Earth. Elsevier, Amsterdam.

Sébillotte M. 1989. Fertilité et systèmes de production. INRA, Paris.

Strullu DG. 1991. Les mycorhizes des arbres et plantes cultivées. Technique et Documentation. Lavoisier, Paris.

Walworth JL, Sumner ME. The diagnosis and recommendation integrated system. Advances in Soil Science, 6: 149-188.

Part II

Biotransformation of Soil

Part Outline

CHAPTER 5

Biotransformation of Carbon and Nitrogen

J. Balesdent

Terrestrial vegetation globally produces 200 billion tons of organic compounds, most of which reach the soil where they are later biodegraded. The elements they contain are released again in the surrounding medium either as nutrients or potential pollutants. Carbon returns to the atmosphere mainly as carbon dioxide. These ongoing flows make up the cycle of organic elements in the biosphere: mainly C, N, P and S.

Man has progressively more intensively and frequently intervened in these cycles at the local and presently at the global level. His intervention may be unintentional or deliberate but regardless, knowledge of these cycles is essential in order to control or maintain them more effectively, to preclude an uncontrollable eventuality or unexpected harmful effects. This knowledge should certainly include the mechanisms involved and, above all, quantification and forecast of flows. As in all complex systems, modelling is the preferred tool for forecasting and even for analysis.

Current Trend: Managing Transitions

Ecosystems and most agrosystems are characterised by a dynamic régime established between the reserve, the flows and environmental conditions.

The perpetuity of these systems evidences the fact that they used to be in a state of equilibrium or, in any case, their deviation was slow. However, man's influence by the end of the twentieth century clearly reveals an increase in change in mode of soil use, either as a consequence of agricultural policies, population increase or simply due to socioeconomic necessity for permanent change. For example, in Western Europe trees are planted and areas left fallow while, concomitantly, new areas are cleared for cultivation. Plants are removed from one plot of land but planted in another. In the tropics, forests are cleared while other areas are abandoned, which become degraded and revert to secondary forests. It is no more a matter of effectively managing a system, but of transitions from one system to another. The system is becoming dynamic. Further, the transport of elements by air and water has made these different ecosystems interdependent: from local and immediate, the study of cycles of elements has become four-dimensional.

DECOMPOSITION OF ORGANIC MATTER

A Series of Enzymatic Reactions

The initial compounds are primarily those of plant origin: in decreasing quantitative order—celluloses, hemicelluloses, pectins, lignins, cutins, proteins, tannins. Next, those of the microflora: microbial polysac-charides, chitins, lipids, lipoproteins, proteins. They are essentially polymerised macromolecules. The first stages of decomposition are generally successive depolymerisations. These are due to specific enzymes, often many of them, which act simultaneously or successively. They are either enzymes of the parent organism, which may be referred to as autolysis, or more often exoenzymes released by the microflora. They eventually become monomers, sugars, amino acids, nitrogenous bases, which are absorbed rapidly by bacterial cells or the hyphae of fungi, with the result that these free compounds are found only in traces in the soil. Lignins are important components of SOM. On the one hand their derivatives are abundant because their biodegradation is slow; furthermore, the latter is only possible if other substrates are present. On the other hand, the free compounds, phenols and quinones are extremely reactive. Many micro-organisms also release metabolites, mainly organic acids, such as lactic, acetic or citric with short chains but considerable reactivity and a strong chelating effect.

In aerobic conditions, the final stage of decomposition is the formation of CO_2. This transformation is called mineralisation of carbon. It is mainly the product of microbial respiration of the soil. Metabolism is identical to that in the respiration of animals. One speaks likewise of soil respiration.

Anaerobic Biotransformations (Table 5.1)

When the diffusion of oxygen is slow and does not meet the requirement of micro-organisms, the conditions may become anaerobic locally and temporarily. This may happen when the soil is waterlogged and very often also in soil not saturated with water within aggregates of sufficiently large size, measurable in centimetres. In a decreasing order of redox potentials, there may be respiration of nitrates and denitrification, reduction of Fe^{3+} and Mn^{4+} as well as fermentation. These reactions are due to facultative anaerobic organisms capable of changing over from aerobic growth to anaerobic. Fermentation converts sugars and proteins to organic acids, especially acetate and butyrate. At still lower potentials there may be reduction of sulphates, methane formation and eventually reduction of protons to molecular hydrogen.

Production of methane in the soil occurs only at a negative redox potential, usually in waterlogged soils that have received a large amount of organics. It so happens typically in flooded rice plantations. The

Table 5 1   *Principal microbial transformations of C and N in soils. The reactions are listed roughly according to a sequence of reducing redox potential in the soil. Oxidation productions are formed, most often from CO_2*

Process		Reaction
Aerobic biotransformations		
Aerobic respiration	$(CH_2O) + O_2$	$\rightarrow CO_2 + H_2O$
Ammonification (example)	$R\text{-}NH_3^+ + H_2O$	$\rightarrow R\text{-}OH + NH_4^+$
Nitrification of ammonium		
nitrition	$NH_4^+ + 3/2\,O_2$	$\rightarrow NO_2^- + 2H^+ + H_2O$
nitration	$NO_2^- + 1/2\,O_2$	$\rightarrow NO_3^-$
Fixation of molecular nitrogen	$N_2 + 8H^+ + 6e^-$	$\rightarrow 2NH_4^+$
Anaerobic biotransformations, possible in partially anaerobic conditions		
Reduction of iron and manganese	$(C\ orga.) + Fe^{3+}, Mn^{4+}$	$\rightarrow Fe^{2+}, Mn^{2+}$
Denitrification	$(C\ orga.) + NO_3^-$	$\rightarrow N_2O, N_2$
Fermentations	$(C\ orga.) \rightarrow$ organic acids, acetate butyrate	
Respiration of nitrates	$(C\ orga.) + NO_3^-$	$\rightarrow NO_2^-$
Obligate anaerobic biotransformations		
Reduction of sulphates	$(C\ orga.\ or\ H_2) + SO_4^{2-} \rightarrow S^{2-}$	
Reduction of CO_2	$H_2 + CO_2$	$\rightarrow CH_4, CH_3CO_2^- + H^+$
Demethylations	$CH_3CO_2^-\ H^+$	$\rightarrow CO_2 + CH_4$
Reduction of proton	fatty acids, $\rightarrow H_2 + CO_2 + CH_3CO_2^-\ H^+$	
	alcohols $+ H^+$	

organisms responsible for this are obligate anaerobic archaebacteria. Methylotrophic bacteria producing methane break up acetic acid into methane and carbon dioxide. At an even lower potential, chemolithotrophic bacteria ensure reduction of CO_2 to CH_2 by molecular hydrogen. Methane transport in the soil is by diffusion and bubbling and especially by vascular transport in hollow rice stems. The presence of oxidising agents, Fe III, nitrates and sulphates, retards its production. Methane formation does not imply a net emission from the soil because soils contain methanotrophic bacteria, which are obligate aerobic, found frequently in the upper layers of methanogenic soils, but also present everywhere. They are also capable of oxidising atmospheric methane.

Since methane is one of the gases responsible for an increase in the greenhouse effect, attempts are presently underway to limit its emission by adopting techniques to reduce its production, such as the composting of rice straw on the surface before ploughing it into the soil.

Effect of Fauna

Microbial colonisation and enzymatic activity are accelerated by the activity of soil fauna which consumes and breaks up plant debris. The most frequently observed groups of saprophytes or detritophages are: termites, ants, oligochaetes, annelids of the earthworm type in neutral soils and enchytreids in more acid soils, and insects such as the Collembola, larvae of Coleoptera and Diptera. The microfauna mainly comprises protozoa, their principal role that of predators, simple saprophytes or plant parasites. All the fauna together seem to play only a negligible part in the decomposition of carbon compared to microflora. On the other hand, some groups are extremely important because of their role as predators or for indirect effects on the activity of micro-organisms. Thus termites and worms improve aeration but appear to aggregate and protect organic matter from decomposition in their casts, helping its contact with clays. Moles, predators of earthworms, and rodents may also modify incorporation of organic matter and water flow.

Various Types of Humus

The fauna that breaks up and incorporates plant debris has an important role in the morphology of humus at the macroscopic level. The classification of forest humus is explained by its variations together with the chemical characteristics of the vegetation and soil. The rapid destruction and ingestion of leaves by earthworms, abundant in eutrophic mediums, lead to formation of humus of the *mull* type, characterised by an almost complete absence of litter and a friable soil structure, due to the aggregation of earthworm casts. In a more acid environment there are no worms; their place is taken by enchytreids and insects. The slower break-up of litter enables formation of several layers characteristic of *moder*. On the soil surface there are several annual layers of recognisable leaves (L layers) and impregnations of fungal hyphae observable on them. Underneath, faecal pellets accumulate in the layer of more decomposed debris, termed F. Humified colloidal matter that has not been incorporated in the soil forms the H layer, immediately above the mineral soil. In highly acid conditions, an accumulation of plant debris occurs, characteristic of *mor*. Biodegradation in submerged soils leads to formation of *anmoor* or, in cold conditions, formation of peat. A conversion of *mull* to *moder* is observed when resinous plants are grown in a sensitive medium; the opposite occurs during fertilisations and limings.

Physical Protection of Organic Matter

Decomposition of organic matter is slower in soil than in water and out of the soil. If soil is stirred, decomposition is accelerated, especially if it is more clayey. The phenomenon occurs also in agriculture during tillage, which explains the decreasing content of organic matter with intensity of tillage in spite of a very high level of plant production and an increase in return of plant material to the soil. From these observations, the concept of physical protection of organic matter has developed. It covers the detention of plant debris in soil aggregates, slow-down in diffusion of oxygen and substrates, adsorption of substrates or enzymes on clay surfaces, and adhesive effect of micro-organisms: all these mechanisms thus render organic matter associated with the mineral matrix less amenable to microbial transformation than free organic matter.

Humic Compounds

Scientists were soon questioned about the presence in soils of complex compounds, coloured, macromolecular, polar, soluble in an alkaline environment. Since these compounds could represent more than half the organic matter in soil, they were called humic compounds: fulvic acids, humic acids and the residue called humin. Interest in the chemical structure of these compounds was considerable because they are extractable and can be analysed. Many theories have been developed to explain their formation and the structure of the 'humic macromolecule'. These products are characterised by a high density of negative carboxylic and phenolic charges (more than 2 milli-equivalents per gram), positive charges of amino groups, and hydrophobic properties. Their average composition is better known now since the introduction of nuclear magnetic resonance of ^{13}C, mass spectrometric pyrolysis and its application to the solid state. They are actually very complex and highly heterogeneous combinations. The constituent elements, the 'elementary building blocks' are derived from sugars, amino acids, lipids, and a wide variety of aromatic com-pounds obtained from plant lignins.

It is not likely that two identical molecules will ever be found. A theory emerges to explain the formation of humic compounds. Because of the constant presence of biodegradable substrates and the microflora which decomposes them, the soil is flooded with enzymes that break the various covalent bonds and affect linkages other than their 'targets'. The transitional products thus formed are likely to recombine with others almost randomly. The macromolecules formed may have a random structure that depends little on the environmental conditions. The reactions could affect biomolecules present in it, such as lignins and proteins, as well as oligomers. Compounds may therefore be formed in solid phase even within inherited plant and microbial structures. As opposed to earlier theories, it appears that humification is both a result of molecular simplification (reduction in size) and progressive increase in size due to condensation and polymerisation.

Functions of Organic Matter

Various types of organic matter are highly reactive and responsible for a number of important functions within the ecosystem and in the environment in general. Apart from the unique role as a reservoir of elements, these types of organic matter constitute one of the major elements of soil fertility, soil aeration and its resistance to degradation and erosion. Their reactive surfaces in some compounds, charged and developing a high potential of van der Waals force, together with their flexibility, provide the soil with cohesive properties. Further, the hydrophobic properties of some compounds slow down the sudden penetration of water in aggregates and check the resultant air pressure that slacks the aggregates. In many poor soils, organic matter is the main component of the absorption complex, that is, the source of charges that retain cations. Besides the properties of ion exchange, organic matter forms complexes or chelates which enable it to fix metals and consequently modify their solubility and prospects for transfer in the soil, helping their utilisation sometimes as trace elements and sometimes manifesting their pollutant characteristics. The macromolecular structure of this type of organic matter and its progressive oxidation helps in trapping potential organic micropollutants. They are the source of energy for a very large variety of heterotrophic micro-organisms: their polymorphism retains a microflora capable of biodegrading a very large variety of organic substrates, including synthetic molecules deliberately introduced by man, such as pesticide molecules, or unintentionally, such as oil products.

NITROGEN CYCLE

The nitrogen cycle is closely linked with the carbon cycle. The common features are microbial and plant biosynthesis, as well as the decomposition of organic matter. However, there are many kinds of biotransformation, associated with a variety of oxidation levels of elemental nitrogen. The general nitrogen cycle covers its exchange between the atmosphere and the biosphere. Almost all transformations take place in the soil, where they form several subcycles (Fig. 5.1).

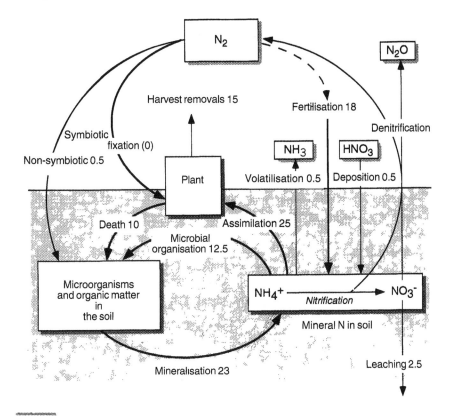

Fig 5.1   Nitrogen cycle in soils. Values corresponding to the probable annual flows are expressed in g N m^{-2} for a cereal crop in northern France.

Fixation of Nitrogen

This is the process by which micro-organisms convert atmospheric gaseous nitrogen to organic nitrogen. It is the natural entry of nitrogen in terrestrial ecosystems. There is a difference between symbiotic and non-symbiotic fixation. Symbiotic fixation may be done by actinomycetes of genus *Frankia*, in association with non-leguminous plants such as alders, *Casuarina* and *Pueraria*. It is mainly by heterotrophic bacteria of genus *Rhizobium* in leguminous plants. Clover, lucerne, lupin, peas, vetch, broom and many other species, in particular tropical ones, often shrubby or woody, such as acacias, are nitrogen-fixing plants. After contact with an absorbent hair on the root, a *Rhizobium*, until then a saprophyte in the soil, induces an enzymatic attack on the cell wall and a localised invasion of the root by bacteria, leading to formation of characteristic nodules in which the *Rhizobium*, previously non-fixing, develop into fixative bacteroids. The bacteroids draw their nutrition from the metabolism of the plant and in return the host plant extracts the formed nitrogen compounds. Artificial inoculation of soils with specific *Rhizobium* is a means of increasing the fixative power of crops. This process is presently in use and there is an increasing demand for the inoculums. Introduction of nitrogen-fixing plants to cut down the inputs has also been the subject of many studies, for example lucerne crops or cereal-legume plant combinations in temperate environments.

Fixation may also be non-symbiotic. This is due to free micro-organisms. Algae of the type Cyanophyceae are involved as are bacteria of genus *Azobacter* or *Azospirillum* in aerobic environments and genus *Clostridium* in anaerobic environments.

While symbiotic fixation, especially by leguminous plants, may involve several hundreds of kg N ha y^{-1}, non-symbiotic fixation generally represents a flow of nitrogen that is much weaker, about 10 kg N ha y^{-1} except in rice fields where it is several tens of kg N ha y^{-1}.

Other Entries of Nitrogen

Other modes of nitrogen entry into ecosystems are through aerial depositions, wet or dry, depending on whether or not they are brought about by precipitation. The sources may be the oxidation of nitrogen by lightning, production of gaseous nitrogen oxides by denitrification or by burning coal and mineral oils, or ammonia from organic wastes, especially effluents from animal farms and fertiliser factories. The predominant form of the deposition is the most oxidised one: nitric acid. This source of nitrogen varies quantitatively in space; it is naturally about several kg N ha y^{-1}. At places it may be somewhat more and even appear as a fortuitous fertiliser. Recent investigations comparing the state of forests in eastern France during the 1970s and 1990s observed over twenty years an enrichment of the environment in mineral and organic nitrogen and also in nitrophilous plants. This effect is manifest mainly on the windward side of these forests.

Lastly, the major form of entry of nitrogen into the biosphere of agricultural soils in developed countries is from the addition of nitrogenous fertilisers, ammonium nitrate or urea, which are themselves synthesised from atmospheric nitrogen. At a global level these entries have increased from 2 Mt N y^{-1} in the 1950s to about 80 Mt today, which represents, however, only half of the biological fixation, evaluated at 160 Mt y^{-1}.

Biodegradation and Mineralisation of Organic Nitrogen

The principal initial forms of organic nitrogen are the proteins of plants and micro-organisms. Next are the amino polysaccharides—constituents of the membrane of fungi or the wall of bacteria, pigments such as chlorophyll and nucleic acids. The nature of decomposition is similar to that of carbon. Organic nitrogen is finally mineralised by various enzymatic reactions, the most common being deaminations of amino acids leading to formation of ammonia. This is known as ammonification. At the level of the entire community of decomposer micro-organisms, the process may be considered an excretion of excess nitrogen from higher organisms.

Nitrification

Bacteria of genus *Nitrosomonas* are capable of oxidising ammonium to nitrates, while bacteria of genus *Nitrobacter* oxidise nitrites to nitrates. This chain reaction, known as nitrification, is the source of energy for organisms that are autotrophic for carbon. Their activity may be affected by many environmental factors. It is strictly aerobic and also reduces with the pH and depends on the availability of mineral carbon, especially in the form of carbonates. In most neutral soils, nitrification is sufficiently rapid so that the ammoniacal form is observed only after the addition of ammonium fertilisers. The balance of charges of the reactions also shows that the reaction releases protons (see Table 5.1). It is one of the principal mechanisms for soil acidification, which leads to recommending the non-use of ammoniacal fertilisers in acid soils.

The two mineral forms are characterised by, among others, a very different mobility. The ammonium ion ion readily binds to the negative charges on the surfaces of the clay-humus complex which is dominant in soils and therefore much less mobile than the nitrate ion which is highly soluble and not retained by the absorption complex. Nitrification inhibitors, synthetic compounds, have also been used to restrict the loss of nitrogen from soils by preserving mineral nitrogen in the ammoniacal form during periods when nitric nitrogen is not needed. Later in the cycle mineral nitrogen, ammonium or nitrates, may be extracted either by the plant cover or by decomposer micro-organisms, thus completing the nitrogen cycle in the soil.

Absorption of Nitrogen by Plants

Mineral nitrogen in the soil is the principal source of nitrogen for higher plants. Although this involves more energy, the nitrate ion is extracted preferentially from the ammonium ion by most plants. Nevertheless, in forest ecosystems the extraction from ammonium through mycorrhizal fungi which assimilate this form may be considerable; extraction is mainly effected by roots located under the layer of litter, where nitrogen is mineralised from dead leaves and roots. The amount of nitrogen needed for a crop is the basis for calculating nitrogen fertilisation. The method adopted in France is that of the forecast of balance of mineral nitrogen, which tries to anticipate the quantity of nitrogen mineralised by the soil. This nitrogen in soil is often half that extracted by a crop, the remainder coming from the fertiliser. The unused part of the fertiliser may in turn be organised by micro-organisms.

Organisation of Nitrogen by Micro-organisms

Mineral nitrogen in the soil is also extracted by decomposer micro-organisms to build up their own cell constituents. This is known as microbial organisation. In this case it is the ammoniacal nitrogen that is preferentially absorbed. Obviously, decomposer micro-organisms may either produce or absorb mineral nitrogen. Decomposition may therefore result in an extraction or release of nitrogen depending on the C/N ratio in the decomposed substrate. Generally microbial immobilisation of nitrogen is observed when the C/N ratio is more than 20-25, while at lower values biodegradation results in release of mineral nitrogen. The level of this threshold value may be explained assuming that the assimilation yield of organic matter by micro-organisms is 30 to 40%, which means that 30 to 40% of the decomposed carbon is used for synthesis of microbial constituents, that the average C/N ratio of the latter is 6 to 8, and that a variable part of the carbon compounds is not biodegradable. It may thus be calculated that for a C/N ratio of the substrate higher than 20-25% the micro-organisms must find a source of nitrogen other than that of the original organic matter. These values of

C/N ratios and assimilation yields are no doubt statistical means and may vary widely depending on the conditions in the medium. Decomposition of plant debris in which the C/N ratio may be 100 is therefore possible only through decomposition elsewhere in the soil of more evolved productions with a lower C/N ratio able to release mineral nitrogen that can be used by the microflora. This cycle is called the mineralisation-reorganisation cycle, which is the most rapid cycle in the soil. Micro-organisms also compete with plants for available nitrogen. In agricultural soils, the ploughing in of harvest residues may thus cause a temporary 'nitrogen hunger' in the crop during the period of growth. It has often been observed that micro-organisms can organise 50 kg N fertiliser ha^{-1} during the growth cycle. This microbial immobilisation may be desirable after a cycle of growth for recycling the excess mineral nitrogen in the organic matter to preclude its leaching.

Another manner of non-biological organisation has also been identified. It involves nitrosation of phenols that may result from a spontaneous reaction of nitrites in highly organic soils.

Volatilisation

Ammonia may also be volatilised. This occurs particularly in hot and dry phases, at a sufficiently high pH, so that the ammoniacal form is not negligible compared to the ammonium form. Volatilisation of ammonia is one of the major losses of nitrogen in regions of intensive animal husbandry and tropical agriculture. It occurs especially during hydrolysis of urea, the principal form of nitrogenous fertiliser used throughout the world.

Leaching of Nitrates and Nitric Pollution

The principal manner in which nitrogen is removed from agrosystems, apart from that in agricultural production, is the carrying of nitrates by water percolating to the base of the soil, called 'leaching'. The nitrates so carried may contaminate phreatic water and make it unfit for human consumption, or contaminate the surface waters and even coastal zones on seashores. In these environments the excess mineral nitrogen, nitric as well as ammoniacal, may trigger a process known as eutrophication in which the availability of fertiliser elements causes a proliferation of algae, followed by their decomposition, which results in a reduction of dissolved oxygen and various phenomena of putrefaction. It has often been said that fertilisers are the cause of pollution in aquifers and especially the water tables. Use of fertilisers labelled with ^{15}N has shown that it is rare for a fertiliser to percolate directly to the water table. Furthermore, the amount of mineral nitrogen in a soil after the harvest is small when fertilisation has been properly regulated. Nitrogen percolates from two sources. One part of the nitrates from the internal cycle of nitrogen is produced by mineralisation of organic matter in the soil at the end of summer and during autumn, because during this period in the Atlantic and Mediterranean climate, temperature and humidity are favourable for mineralisation, there is no extraction by the plants and rainfall is heavy. The other part, i.e., excess of inputs leading to accumulation of residual mineral nitrogen may also percolate. These inputs may be fertilisers, as well as effluents from animal farms, other sources of organic matter or fixation of nitrogen by leguminous plants. Another cause of nitric pollution is intense mineralisation of the organic matter reserve in the soil as a result of stress when this is not taken into account by reducing fertilisation. Planting of crops on grasslands or forest soils, intensification of tillage, reduction in restoration of organic matter to the soil, and vegetable crops present a high risk of pollution. In short, any change in the established regime between inputs and removal of organic matter from the soil alters the input-removal balance of nitrogen. Thus, policies to limit nitric pollution must take into account many technical improvements and incentives when dealing with changes in the nitrogen cycle of the soil.

Denitrification, Pollution Control or Pollution?

When an assessment is made of the inputs and removals of nitrogen in a region or a country like France, it is observed that the known removals such as discharge of nitrates into the sea and removal of proteins, represent less than half of the inputs: fertilisers, atmospheric precipitation and biological fixation. The difference could be the increase in nitrogen reserve of the soils and even in subterranean water tables. However, the most likely mechanism to explain this is denitrification, diffused or confined to certain milieus.

Denitrification is conversion of nitrogen from nitrates to gaseous nitrogen (N_2) and nitrous oxide (N_2O), also called nitrogen protoxide or laughing gas. It may be produced by a population of facultative anaerobic

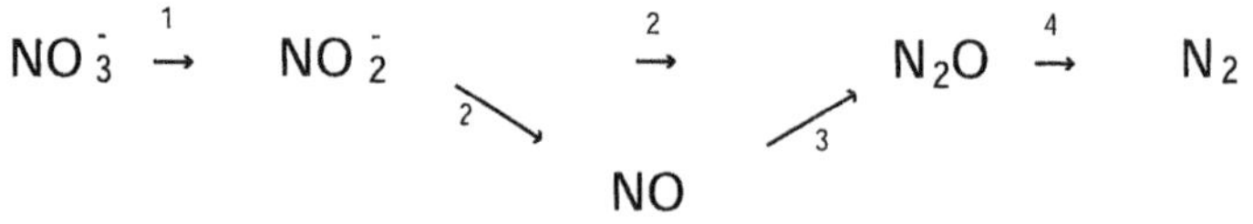

$$NO_3^- \xrightarrow{1} NO_2^- \xrightarrow[2]{} \xrightarrow{2} \xrightarrow[3]{} N_2O \xrightarrow{4} N_2$$

$$NO$$

1 Nitrate reductase   2 Nitrite reductase   3 Nitric oxide reductase   4 Nitrous oxide reductase

Fig 5.2 Enzymatic sequence of biological denitrification.

bacteria, representing a few percentiles of the total bacterial population in the soil. The nitrate ion is reduced by a series of reductases according to the sequence of reactions shown in Fig. 2 during which nitrogenous products are used as electron acceptors for respiration, playing the role of oxygen that they replace in its absence. The process requires anaerobic conditions and the presence of assimilable carbon, an energising substrate and an electron acceptor. This is one of the nitrogen flows that is most difficult to measure *in situ*. Contact of soil with acetylene which inhibits conversion of N_2O to N_2, followed by measurements of N_2O emissions by chromatographic dosages in the gaseous phase, is the most commonly used method. The natural abundance of ^{15}N in nitrate, which increases during denitrification and leads to enrichment of the residual nitric nitrogen reserves is also a preferred way of showing this transformation. It was thus possible to show that it is produced typically in certain organic environments, such as naturally waterlogged zones, river banks or river-bank works. These areas represent the transition of surface waters from terraces or plateaus at a higher level and may dissipate high concentrations of nitrate, which are often observed at the base of some agricultural catchment areas. They have also been found to be genuine natural purifiers of catchment areas even if they have not been drained for sanitary reasons. Biological denitrification, induced or provoked, is probably the best way of removing nitrates from waters. Nevertheless, this transformation not only has beneficial effects because alongside N_2, it also produces N_2O—a gas with a high radiative absorptivity and a long life, considered to be the fourth contributor after CO_2, chlorofluorocarbons and methane, to the increase in greenhouse effect and its concentration in the atmosphere is presently increasing at the rate of 0.25% per annum.

Other organisms, obligate or facultative, are capable of reducing ammonium to nitrates, a reduction termed dissimulative as opposed to assimilative. The flows involved in the soil are negligible, however, except in soils with a very high content of organic matter and likely to receive water charged with nitrates. There are also non-enzymatic mechanisms which lead to production of oxides of nitrogen and N_2. These are in particular the reduction of nitrates and nitric acid, sometimes called chemical denitrification.

METHODS OF INVESTIGATION

There are many methods for studying the C and N cycles at various scales, from the availability of genetic labelling of bacteria involved in particular transformations to the study of catchment areas equipped for measuring flows. Nonetheless, isotopic tracers are the preferred method as they enable tracking within complex populations the various constituents under study, atoms from a particular source and their rates of transformation in a continuous regime. A distinction must be made between artificial tags and natural isotopic tracers.

Compounds artificially enriched with ^{14}C (radioactive isotope) and ^{15}N (stable isotope) have been used since 1960 to track developments in plant compounds and nitrogenous fertilisers. The radioactive ^{13}N isotope, with a very short half-life, enables effective labelling in very small amounts and has been used for tracking some stages in the metabolism of nitrogen and in particular denitrification. Although such trackings are very useful, there are two major limitations. Incorporation of the tracer is generally not done under the natural conditions of the element. Furthermore, the given element can only be tracked for the duration of the experiment. That is why laboratory tracking is now mostly supplemented by information collected from variations in the natural abundance of stable isotopes, notably ^{13}C and ^{15}N.

Variations in the natural abundance of stable isotopes are limited to, at most, a few percentiles in relative value. They can be used to reveal processes of transformation: denitrification and volatilisation of ammonia are processes that provoke isotopic fractionation, i.e., a small difference in behaviour between heavy and light isotopes. These transformations can be shown by the natural abundance of ^{15}N.

This natural abundance can also help in identifying the origin of the compounds studied. The abundance of ^{13}C has been used, for example, since the end of the 1980s as a tracer of carbon dynamics. The method is based on the premise that plants with a photosynthetic cycle in C4 (mainly tropical graminaceous plants and in Western Europe, maize) are naturally richer in ^{13}C than plants with a photosynthetic cycle in C3. The phenomenon results from the difference in fractionation associated with the reaction of primary carboxylation. When vegetation of the C3 type is substituted with that of the C4 type, the new organic matter of C4 origin will progressively replace the former in the soil. As organic matter in the soil preserves the isotopic composition of the previous vegetation, it is possible to measure the rate of its replacement by following the rate of change in composition in ^{13}C of the soil. We thus measure the turnover rate of organic matter. As any individual organic fraction can be measured, different pathways of transformation can be tracked and that too quantitatively. Monocultures of maize in France, afforestation of grasslands or the use of forests as grazing areas in the tropics, have been explored by this method for measuring the distribution of residence times of carbon in soils (Fig. 5.3).

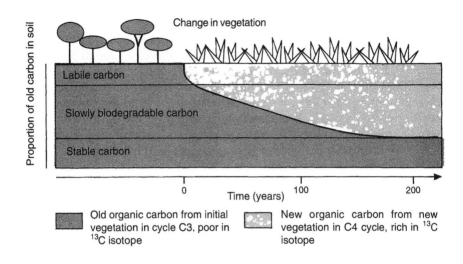

Proportion of old carbon in soil

Change in vegetation

Labile carbon

Slowly biodegradable carbon

Stable carbon

0   Time (years)   100   200

Old organic carbon from initial vegetation in cycle C3, poor in ^{13}C isotope

New organic carbon from new vegetation in C4 cycle, rich in ^{13}C isotope

Fig. 5.3 Example of utilisation of naturally abundant isotopes.

[14]C dating of organic matter in soil has also been used to show the existence of biologically stable carbon, thousands of years old, in most soils. Further, release of ^{14}C by thermonuclear aerial tests from 1955 to 1964 has fortuitously labelled carbon of the entire biosphere. In 1963, its content in the atmosphere was double the norm. Sequential datings have helped measure the progressive incorporation of carbon in soils.

QUANTIFICATION

Fundamental Variables and Spatial Distribution of Various Biological Functions

Modelling cycles of elements should take into account the nature of the processes involved, especially their amplitude, rate, and variations in this rate depending on environmental factors. The fundamental factors for rate of decomposition of organic matter can be listed according to their relative contribution to the variation of this rate in the natural environment.

First among these factors is soil temperature. The activity of micro-organisms increases exponentially with temperature, increasing the rates of decomposition by 3 orders for every increment of 10°C. In the tropics decomposition of organic matter is therefore nominally fourfold higher than in the northern region of France. The second factor is soil moisture.

Here, decomposition correlates empirically with no simple univocal relationships established. It is generally accepted that decomposition is maximal at retention capacity. It reduces when the moisture potential increases and can thus be divided by a factor of 4 when the pF increases from 3 to 4.2, the value corresponding to wilting point. It should be kept in mind, however, that some microflora have adapted to dry conditions. Under humid conditions, a slow-down in rates, explained by reduced availability of oxygen, is rather difficult to describe by a universal law. Temperature and humidity are most important for distribution of reserves and flows of carbon on the planet. The net primary production and hence flows of organic matter reaching the soil depend mainly on precipitation, subsequently corrected by evapotranspiration. Since the rates of decomposition depend primarily on temperature, it should be noted that schematically the largest carbon reserves are found in a cold and humid climate and the smallest in a hot and dry climate. On dividing the planet into large biomes, it was possible to integrate these reserves and flows. Thirdly, soil texture and tillage regulate physical protection. The physicochemical properties of soil such as pH, base saturation and calcium carbonate content, do not affect the rates of decomposition in a simple and univocal manner.

Soil classification, incorporating physicochemical factors, may provide a more acceptable typology. In fact, soil properties resulting from the parent rock also indirectly affect the carbon cycle. Effects of fauna have already been described. The decomposition of various types of organic matter competes with other non-biological transformations. Their dissolution and subsequent transport by water are an example of this. Thus, in cold conditions and where drainage is effective, organic acids may escape mineralisation and be carried to depth. Their ability to carry Fe and Al cations is responsible for the process of podzolisation. *Podzols* are thus found in the taiga zone and in mountains; they are also observed on some sandy substrates in zones with temperate and equatorial climates.

Adsorption is also a mode of protection against decomposition, common under cold conditions with simultaneous pronounced seasonal desiccation. Such conditions exist in continental climates and in soil saturated with a sufficient quantity of clay. Organic matter protected in this way is likely to accumulate in large quantities. The soil is *chernozem* (Russian for black earth), belonging to the class of isohumic soils according to the French classification. There is another class of soils in which the rates of decomposition of organic matter are extremely slow. These soils have developed from volcanic material in climates that are always humid and belong to the class of Andosols. The absence of desiccation in them enables formation and preservation of secondary aluminosilicates known as allophanes, derived mainly from volcanic glass. Allophanes form veritable gels, developing surfaces and a high density of cationic charges which impart to the soil an apparent density that may be less than 0.3 g cm^{-3}. The organic matter is intimately mixed, adsorbed, and the decomposition here may be 5-10 times slower than in a ferrallitic soil in an identical climate.

Modelling

Many mathematical models are presently used in research to simulate the decomposition of organic matter in soils. Details vary depending on the required objective—to enhance knowledge, to forecast carbon or nitrogen, or to simulate a season or a century.

Figure 5.4 presents a model simulating the carbon and nitrogen cycle in soil. Organic carbon and nitrogen are split into separate compartments. The definition of these compartments is mathematical; separation of types of organic matter depends on their lifespan and the flows which link them, covering concomitantly naturalistic concepts, either functional (e.g., soil decomposers) or chemical (e.g., lignins, mineral nitrogen in the soil). Each compartment follows a first-order kinetics of decomposition. The rate constants are affected by a multiplicative factor depending on environmental factors: temperature, humidity, soil texture and tillage, the last two accounting for physical protection. The first three compartments are supplied by dead plant and root depositions. The products of decompositions of each compartment form CO_2, microbial constituents and slowly decomposing organic matter in proportions which depend on the organic substrate and the environment. Soil also contains organic matter that has been stable for centuries. Not all these compartments can be isolated. Only structural polysaccharides and lignins can be separated. There are also indicators of the microbial biomass in the soil, but this does not always correspond to the active biomass. The nitrogen cycle is described mainly by mineralisation of each compartment at the same rate as that of carbon and by microbial organisation together with decomposition of celluloses and lignins. Leaching of nitrates and denitrification depend on the state and circulation of water in the soil. These models may be combined with simulation models for the growth of plant populations and circulation of water and solutes.

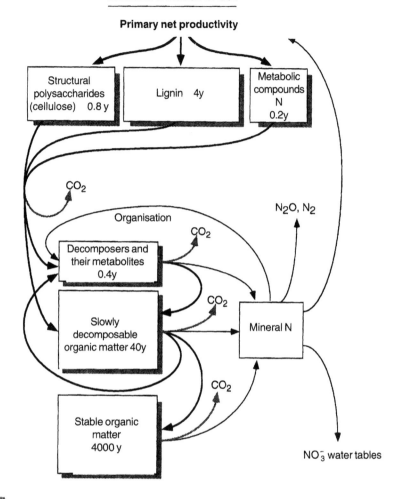

Fig. 5.4 Structure of a simulation model of the C and N cycles in soil. The average times of residence of carbon in each compartment are those for fictitious silty soil in a temperate forest.

Glossary

Actinomycetes	Micro-organisms, often filamentous, classified as bacteria, with a morphology intermediate between that of fungi and bacteria. They are an important part of the microbial population of soils.
Aerobic	Where molecular oxygen (O_2) is present. Refers to organisms using oxygen for respiration.
Anaerobic	Where there is no molecular oxygen (O_2). Refers to organisms that can develop without oxygen.

Biodegradation (of organic compounds)	Change due to action of living organisms, mainly micro-organisms that draw energy from those compounds, which may extend to mineralisation but may also leave intermediate compounds in the milieu.
Denitrification	Reduction of nitrate or nitrite ion to molecular nitrogen (N_2) or to nitrogen oxides.
Eutrophic	Refers to a medium rich in nutritive elements.
Humic (humic compounds)	Organic compounds resulting from humidification, often macromolecular, with a complex chemical and random structure.
Humiferous soil	Type of soil rich in organic matter.
Humification	Chemical transformation in soil of compounds and living organisms that imparts a molecular structure differing from that of live molecules. Enzymatic reactions, recombinations of free radicals and spontaneous reactions are responsible for the process.
Methanogen	Methane-producing micro-organism.
Methanotroph	Organism feeding on methane.
Mineralisation	Transformation of organic forms of C, N, P, S into inorganic forms (HCO_3^-, CO_2, NH_4^+, PO_4^{3+}, etc.).
Nitrification	Transformation of ammonium ion into nitrate ion. It is accomplished by bacteria.
Nitrogen fixation	Conversion of atmospheric molecular nitrogen (N_2) to organic nitrogen by micro-organisms.
Oligotroph	Said of an environment poor in nutritive elements.
Organic fraction	Part of organic matter identified either by a method of separation or dosage or by a particular functional characterisitic.
Organisation of N, P, S	Conversion of inorganic forms (nitrate, ammonium, phosphate, sulphate, etc.) into organic forms.
Rhizodeposition	Transfer of organic components of roots to the soil by exudates of soluble metabolites through senescence or death.

Note: Other scientific terms are defined at the place of their appearance in the text.

FURTHER READING

Lal, R. Follett Ronald F. (eds.) 1997. Soil Processes and the Carbon Cycle. Lewis Publishers, Inc.; ISBN: 0849374413, 624 pp.

Paul EA, Clark FE (eds.). 1996. Soil Microbiology and Biochemistry. Academic Press; ISBN: 0125468067; 340 pp. (2nd ed.).

Stevenson F-J. 1982. Nitrogen in Agricultural Soils. Amer. Soc. Agron., Madison, WI, 940 pp.

Stevenson F-J. Cole MA. 1999. Cycles of Soils: Carbon, Nitrogen, Phosphorus, Sulfur, Micronutrients. John Wiley & Sons; ISBN: 0471320714, 448 pp. (2nd ed.).

CHAPTER 6

Micro-organisms in the Transformation of Minerals: Effect on Formation, Functioning and Development of Soils

J. Berthelin

Ever since life appeared on the Earth 3.5-4 billion years ago, with the first lower unicellular organisms, ancestors of actual bacteria, micro-organisms (or protists) have colonised and grown in all the super-ficial terrestrial environment (soils, continental waters, oceans, sediments, atmosphere...). They have established themselves by adopting nutritive and energetic strategies, as shown in Figure 6.1, which are highly diverse and enable them to exist by using organic compounds (micro-organisms known as chemo-organotrophs or heterotrophs) as well as mineral compounds (bacteria known as chemolithotrophs or autotrophs). They grow and reproduce in the presence or absence of oxygen (aerobic or anaerobic respiration and fermentation).

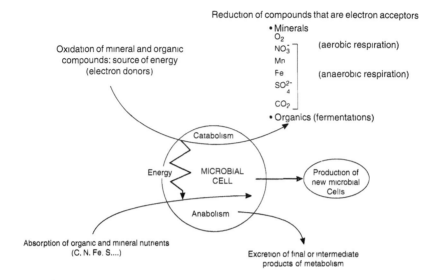

Fig. 6.1 General simplified diagram of functioning of chemolithotrophic and chemo-organotrophic micro-organisms.

They have also developed mechanisms of adaptation and resistance to the very numerous pressures of their environment (nutritional deficiencies, drought, extreme temperatures, acidity, salinity, stress).

Among the main natural environments, soils, veritable reservoirs of micro-organisms, contain all known groups of lower unicellular protists (bacteria, actinomycetes, cyanobacteria) and higher protists (fungi, algae, protozoa). These micro-organisms (or protists) are not distributed throughout the entire thickness (or profile) of the soil but prefer sites or niches where nutritive and energetic conditions are satisfactory, such as surface litter, humus and surface soil horizons (or layers) or the root zone (rhizosphere).

They are divided into groups, families, genera and species, as well as into communities or physiological groups comprising various populations with one or several well-defined metabolic activities. Thus among the fungi and bacteria which are qualitatively and quantitatively the two most important groups, many communities may be identified that have the same nutritional requirements or the same metabolic abilities (cellulosis, lignolysis, amylolysis, nitrogen symbiotic fixation, ammonification...). Some of these metabolic activities, for example nitrification (or oxidation of ammonia to nitric acid), sulpho-oxidation (or oxidation of sulphides, elemental sulphur, thiosulphates to sulphuric acid), oxidation and reduction of manganese and iron, solubilisation of phosphates, silicates and carbonates are, or may be, directly or indirectly involved in dissolving or causing precipitation of mineral elements, weathering and degradation of minerals, and formation of deposits of various metallic elements.

Soil bacteria and fungi therefore appear to be effective agents in the degradation, transformation and formation of minerals or mineral constituents of soils because they can develop and promote mechanisms for dissolution and insolubilisation of mineral elements.

DISSOLUTION AND WEATHERING OF MINERALS

Various processes in the dissolution of mineral elements leading to a relatively simple transformation of primary minerals, such as micas, to secondary minerals like clays, to a total breakdown of minerals, involve soil bacteria and fungi.

Production of Acids and Complexing Substances

During the course of biodegradation of organic matter essentially of plant origin (leaf litter, root exudates, debris etc.) which they use as a source of carbon and energy, chemo-organotrophic (or heterotrophic) bacteria and fungi (all heterotrophic) produce acid compounds in the soil according to the simplified eqn (1).

$$\text{Organic matter } (CH_2O)_n \rightarrow \begin{cases} \text{Organic acid (RCOOH)} \\ CO_2 \rightarrow H_2CO_3 \text{ (carbonic acid)} \end{cases} \quad (1)$$

These acids (carbonic, carboxylic-aliphatic or phenolic etc.) excreted in the soil may occur in the solubilisation of mineral elements as shown in the general eqn (2):

$\text{(Metal)}^+ \text{ (Mineral anion)}^- + \text{Acid } (H^+ R^-) \rightarrow H^+ \text{ (mineral anion)}^- + \quad (2)$
$\text{(Metal)}^+ R^-$

where

(Mineral anion)$^-$ = silicates, phosphates, carbonates, sulphides,...
and $R^- = NO_3^-, SO_4^{2-}, HCO_3^-, RCOO^-$.

As an example, the simplified reaction (3) indicates the transformation of insoluble tricalcium phosphate to monocalcium phosphate which is much more soluble because of the effect of carboxylic acid:

$$Ca_3(PO_4)_2 + R (COOH)_2 \rightarrow R (COO)_2 Ca + 2Ca(PO_4H_2)_2 \quad (3)$$

Figure 6.2 shows in a milieu containing particles of insoluble tricalcium phosphates (white dots), a dissolution halo (absence of white grains of the mineral) around a colony of *Enterobacter agglomerans*

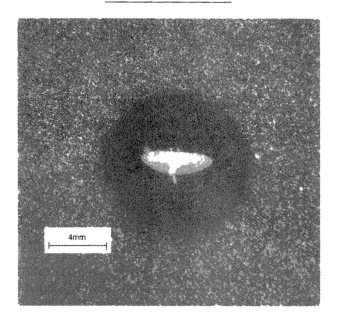

Fig. 6 2   *Enterobacter agglomerans* bacterium solubilising tricalcium phosphates. A dissolution halo of phosphate grains appears around the bacterial colony.

bacterium isolated from the rhizosphere of maize. This bacterium, by producing acids, has dissolved mineral particles in its new environment. In soils, and especially in the rhizosphere of plants, there are usually a hundred thousand to ten million bacteria per gram of dry soil that can dissolve insoluble or weakly soluble phosphates of calcium due to the production of acid compounds.

Such production of acids in both aerobic and anaerobic conditions is involved in the dissolution and weathering of various minerals: phosphates, carbonates, silicates, oxides, sulphides.

Some micro-organisms and certain conditions of the surroundings (especially aerobic) promote production of more efficient substances such as oxalic and citric acids, with a high complexation effect on mineral elements (iron, aluminium, copper, zinc, lead, uranium). Several compounds such as siderophores with trihydroxamic or tridiorthophenolic functional groups have, vis-a-vis ferric iron, a more specific complexation effect. To these numerous productions of organic acids and carbon dioxide (that can result in carbonic acid) may be added the production of nitric acid by nitrifying chemolithotrophic bacteria (*Nitrosomonas* and *Nitrobacter*) during oxidation of ammonia and of sulphuric acid by *Thiobacilli* during oxidation of sulphides, sulphur and thiosulphates.

Phenomena of Oxidation and Reduction of Mineral Compounds

Some soil bacteria may be responsible for the phenomena of dissolution, using redox reactions that specifically involve minerals which they use either as electron donors (energy source) or as electron acceptors. This applies to iron-reducing bacteria and *Thiobacilli*. The bacteria known as iron-reducers (*Bacillus, Clostridium* etc.), which exist in aerobic-anaerobic conditions (in the presence or absence of oxygen) or strictly anaerobic (in the absence of oxygen) may in the absence of oxygen, manganese and, for some of nitrate, and during fermentation or anaerobic respiration, reduce ferric to ferrous iron as shown in eqns (4) and (5):

$$\text{Source of carbon and energy (fermentable or not)} \tag{4}$$
$$\rightarrow n e^- + H^+ \pm \text{products of fermentation}$$
$$Fe(OH)_3 + 3H^+ + e^- \rightarrow Fe^{2+} + 3H_2O \tag{5}$$

By this mechanism, they dissolve oxyhydroxides of ferric iron (goethite hematite), which are minerals of very low solubility.

Thiobacilli, in particular *Thiobacillus ferrooxidans* and *Thiobacillus thiooxidans*, which are acidophilic chemolithotrophic bacteria, oxidise reduced compounds of sulphur, including sulphides. One of them (*Thiobacillus ferrooxidans*) oxidises ferrous iron. These bacteria can therefore dissolve sulphide minerals (pyrite, chalcopyrite, arseno-pyrite...) by forming soluble sulphates according to the simplified eqn (6):

$$FeS_2 + \frac{7}{2}O_2 + H_2O \rightarrow FeSO_4 + H_2SO_4 \tag{6}$$

But, by producing sulphuric acid and ferric sulphate (for *Thiobacillus ferrooxidans*), they can with this strong acid and powerful oxidiser, act on other minerals (oxides, carbonates, phosphates, silicates) simultaneously with dissolution of sulphides. Figure 6.3 shows a pyrite particle corroded by *Thiobacillus ferrooxidans*.

INSOLUBILISATION, ACCUMULATION AND DEPOSITS OF MINERAL ELEMENTS

Separately or in competition with the processes of solubilisation, chemo-organotrophic and chemolithotrophic microorganisms are involved in the processes of insolubilisation associated with reactions of oxidation, reduction, bioaccumulation of mineral elements and the biodegradation of soluble organomineral compounds.

Processes Involving Redox Mechanisms

The insolubilisation of mineral elements by redox phenomena mainly involves, on the one hand, oxidation of iron and manganese which can

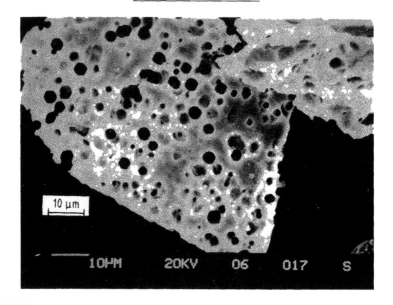

Fig 6 3   Pyrite (iron sulphide) weathered by *Thiobacillus ferrooxidans* bacterium.
Dissolution pits are clearly visible on the pyrite.

then precipitate in the form of hydroxides and, on the other, reduction of
sulphates to sulphides that may later, combined with transition metals,
precipitate in the form of insoluble metal sulphides.

In an aerated acid environment, *Thiobacillus ferrooxidans* oxidises iron
which remains in solution if the medium is sufficiently acidic (pH 1.5 to
2.0). Ferric iron will precipitate if acidity decreases.

There are many bacteria with specific structures known as sheaths
(*Leptothrix*) or stalks (*Gallionella*,....) or as budding bacteria which may, in
slightly acid or neutral, aerated or microaerophilic conditions oxidise
both these metals by forming hydroxides or oxyhydroxides. For example,
with iron, these bacteria promote formation of rust deposits that appear
on water puddles, or at the emergence of aquifers on the surface of soils,
or in agricultural drains during iron ochre deposits.

Figure 6.4 shows cylindrical sheaths formed by ferric iron. These
sheaths were produced by bacteria of genus *Leptothrix* which have moved
away.

Equations (7) and (8) summarise this oxidation (eqn. 7) leading to the
deposition of hydroxides and insoluble iron (eqn. 8).

$$Fe^{2+} + \frac{1}{4} O_2 + H^+ \rightarrow Fe^{3+} + \frac{1}{2} H_2O \qquad (7)$$

$$Fe^{3+} + 3OH^- \rightarrow Fe(OH)_3 \qquad (8)$$

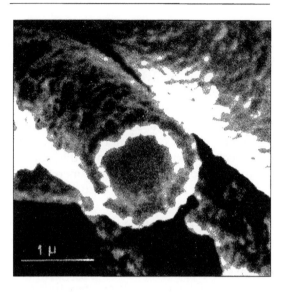

Fig 6 4   Sheaths of ferric iron deposits formed by *Leptothrix* sp. bacterium.

In environmental conditions without oxygen (strict anaerobiosis), sulphates may be reduced by bacteria known as sulphate reducers which use products resulting from fermentation (lactic acid, acetic acid, ethanol, hydrogen, ...) as a source of energy (electron donors) and sulphates as electron acceptors in anaerobic respiration. Equation (9) summarises this process:

$$2CH_3 CHOHCOOH + SO_4^{2-} \rightarrow 2CH_3COOH + 2CO_2 + S^{2-} + 2H_2O \quad (9)$$

These sulphate-reducing reactions play an important role in deposits of metal sulphides and may be at the origin of sulphide ores.

Process Involving Organomineral Compounds: Biodegradation of Organometallic Complexes and Bioaccumulation of Metals

Processes of insolubilisation and concentration of mineral elements may involve biodegradation of soluble organometallic complexes comprising one or several 'ligands' or organic complexing anions (for example, hydroxycarboxylic acids, amino acids, hydroxamic acids) and one or several metals combined with them. Micro-organisms can use the organic part of the complex as a source of carbon and energy and thereby release the metal which, depending on the environmental conditions (acidity, redox potential, ionic concentration), may precipitate.

These micro-organisms may also be involved in phenomena of accumulation and concentration of metals that correspond either to the absorption of metallic elements in the microbial cells far in excess of nutritive requirements or their adsorption, or complexation by the cell wall constituents.

Such an accumulation involves living or dead micro-organisms, which leads to a distinction between bioaccumulation by living cells and biosorption by dead cells. Bioaccumulation may correspond to a resistance mechanism to the metals as, for example, the production of proteins that preferentially fix cadmium or copper or the production of exocellular polysaccharides immobilising the metals, or even a modification in the composition of bacterial membranes with a higher production of dicarboxylic amino acids, ensuring better immobilisation of cadmium.

Microbial biosorption of metals is therefore defined as a process of adsorption or complexation by dead cells and the cellular constituents, also called biosorbents. It corresponds to fixation of metal ions by various functional chemical groups present in the cell wall, such as $COOH - NH_2$, $- SH, - OH, - H_2PO_4$, ... Bacteria and fungi may thus accumulate metals up to 30% of their dry matter weight.

EXAMPLES OF THE EFFECT OF SOIL FUNCTIONING PROCESSES

In soils, microbial activity depends primarily on nutritional and energetic conditions and secondly on various factors such as humidity and water saturation of soils, temperature, acidity and alkalinity, salinity,... . Parameters such as the nature of vegetation, parent rock, hydric conditions etc. will also play a fundamental role.

Figure 6.5 summarises the microbial mechanisms of dissolution/ weathering of minerals and the conditions under which they occur.

In acid brown forest soils associated with fir and beech forests with a herbaceous cover of graminaceous plants, on granitic or sandstone rocks as found in north-eastern France, the rapid and effective biodegradation of plant material and residue lead to good mineralisation (decomposition to CO^2, H_2O, NO_3) of plant organic matter. In these environments, the activity of nitrifying bacteria ensures production of abundant amount of nitric acid even during the winter season. The acid formed in this manner is then responsible for the dissolution of mineral elements, preferably those that are alkaline, and alkaline earths (Ca^{2+}, Mg^{2+}, K^+) when the parent rock has a large mineral reserve (granites with feldspar, plagioclases, phyllosilicates, ferromagnesium) or even aluminium when the parent rock has a limited mineral reserve, as in some sandstone areas in the Vosges mountains. Apart from the effect on weathering of minerals, this aluminium in the ionic form may have a toxic effect on vegetation and on freshwater fauna if its content becomes relatively high in the soil solution.

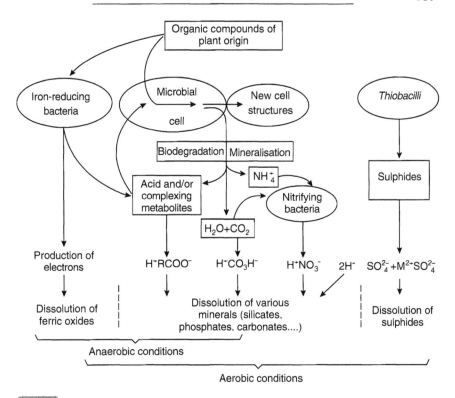

Fig 6 5 Involvement of soil micro-organisms in the dissolution and weathering of minerals.

In soils such as podzols, the more limited activity of the microflora causes a decrease in the biodegradation of organic substances such as those with a high complexing effect, which can thus as a consquence dissolve elements such as iron and aluminium from the surface horizons and transport them to horizons of accumulation at the base of the soil profile (illuvial horizons). Further, in the very same soils, micro-organisms participate also in the production of organic acidic compounds which can form soluble organometallic complexes with metallic elements from the rocks.

The effects of permanent or temporary waterlogging (hydromorphic soils) or even a momentary saturation of soil pores with water, promote phenomena such as bacterial reduction of manganese and iron, as well as production of aliphatic acids resulting from fermentations. This bacterial reduction helps in dissolving oxyhydroxides of iron, which is well observed in the white and green coloration of certain horizons. This considerable mobilisation of iron in the ionic ferrous form provides energy sources for ferro-oxidising bacteria (*Leptothrix, Gallionella*) that develop at the level of emergence of aquifers or in drains and which they subsequently block (ferric iron ochre deposit).

In the rhizosphere of plants, micro-organisms which live on the roots or immediately around them and the mycorrhizae (symbiotic associations of fungi and roots) have a very significant effect on the dissolution of mineral elements (phosphorus, potassium, iron, magnesium,...) from rocks and minerals, and help to improve the mineral nutrition of plants. This rhizospheric microbial activity plays a role not only in the mobility and bioavailability of major elements, but also of trace elements (oligoelements, heavy metals). For example, after one or two years in the rhizosphere of a plant, or even after only a few months under certain experimental conditions, weathering of mica (phlogopite or biotite) to clay minerals of the vermuculite type can be observed.

These few examples, fully detailed in the works cited in the references, help to establish the fundamental role played by micro-organisms in weathering and transformation of minerals in soils, and mobility and availability of mineral elements for various organisms, in particular plants, and hence the functioning of 'soil-plant' systems.

Other effects are evident in the extraction and concentration of metals. Industrial applications have already been adopted for treating ores, industrial minerals and wastes. They hold promise for fundamental studies and their application.

FURTHER READING

Berthelin J. 1987. Des bactéries pour extraire des métaux. La Recherche, no. spécial 'L' avenir des biotechnologies', 188: 720-725.

Berthelin J, Leyval C, Toutain F. 1994. Biologie des Sols. Rôle des organismes dans l'altération et l'humification. In: Pédologie, Tome 2. Constituants et propriétés du sol. M. Bonneau, B. Souchier (eds.), pp. 143-237. Masson, Paris.

Berthelin J, Munier-Lamy C, Wedraogo FX, Belgy G, Bonne M. 1990. Mécanismes microbiens d'acidification et d'altération intervenant dans les sols bruns acides et les podzols forestiers. Science du Sol, 28: 301-318.

Deneux-Mustin S, Rouiller J, Durecu S, Munier-Lamy C, Berthelin J. 1994. Détermination de la capacité de fixation des métaux par les biomasses microbiennes des sols, des eaux et des sédiments: intérêt de la méthode du titrage potentiométrique. C.R. Acad. Sci. Paris, Série II: 1057-1062.

Houot S, Berthelin J. 1992. Submicroscopic studies of iron deposits occurring in field drains: formation and evolution. Geoderma, 52: 209-222.

Leyval C, Berthelin J. 1991. Weathering of a mica by roots and rhizospheric microorganisms of pine. Soil Sci. Soc. Amer. J., 55: 1009-1016.

Mustin C, Berthelin J, Marion P, de Donato P. 1992. Corrosion and electrochemical oxidation of a pyrite by *Thiobacillus ferrooxidans*. Applied and Environmental Microbiology, 58: 1175-1182.

Watteau F, Berthelin J. 1994. Microbial dissolution of iron from minerals: efficiency and specificity of hydroxamate siderophore in the dissolution of ferric oxides. Eur. J. Soil Biol., XXX: 1-9.

CHAPTER 7

The Rhizosphere

P. Lemanceau, T. Heulin

The term rhizosphere (Gk. *rhiza:* root, *sphaira:* that which surrounds it) was first proposed in 1904 by an Austrian scientist (Hiltner) to describe the **zone of soil that surrounds the root** and which is **directly or indirectly influenced by the root**. The rhizosphere can be considered as being the half-hidden part of the root system which itself is concealed. Within the rhizosphere there is the **rhizoplane** which corresponds to the soil-root interface, and the soil **which adheres to the root system**, i.e., the soil that remains attached to the root after vigorous shaking. The root modifies the physicochemical and microbiological properties of rhizospheric soil. These modifications determine the so-called 'rhizospheric effect'. This effect results from the extraction of water and mineral elements by the root, but overall from the release of organic compounds. The volume of soil subjected to the rhizospheric effect is determined by the zone of diffusion of soluble organic molecules and volatile compounds released by the root.

Methodological Difficulties in Studying the Rhizosphere

Estimation of the rhizospheric soil volume is difficult. In fact, the definition of rhizospheric soil depends on the technique adopted to separate this soil from non-rhizospheric soil. Analytical results from this zone have therefore no absolute value but rather help to compare within a given study the effect of different experimental conditions.

In spite of the difficulties encountered in its study, the rhizosphere has been investigated by various scientific disciplines. The characteristics of the rhizosphere influence plant growth and health. For physiologists, the rhizosphere is a zone of ion exchanges, competition for oxygen and release of organic compounds. For soil scientists, the rhizosphere is a zone of soil in which exchanges and water diffusion are modified. For microbiologists, the rhizosphere is a zone of soil in which the microbial density and acitivity are stimulated by the release of organic compounds. Lastly, for plant pathologists, the rhizosphere is a zone of soil in which saprophytic growth of pathogenic micro-organisms is also stimulated by the release of organic compounds.

The synthesis of the applied methodologies and the results obtained in these different disciplines contribute to our knowledge of this difficult-to-access zone. The progress achieved allows one to envisage management of the biotic and abiotic interactions occurring within the rhizosphere in order to limit the use of pesticides and synthetic fertilisers for a more environment friendly agriculture.

A VERY SPECIAL ENVIRONMENT

Rhizospheric soil shows physicochemical properties that differ from those of non-rhizospheric soil. The rhizosphere is actually a place where intensive exchanges occur between the soil, the roots and the microflora. These exchanges are numerous, such as release of organic compounds and ions, absorption of water and ions by the root, respiration of the root and microflora, and synthesis of various microbial metabolites. Altogether these exchanges determine an increase of the organic matter content and a modification in the ionic is and gaseous balances within the rhizosphere.

Rhizodeposition is Responsible for Formation of Rhizospheric Soil

The rhizospheric effect is largely determined by the release of various organic compounds grouped under the term 'rhizodeposition'. More generally, the release of a part of the photosynthetates in the soil contributes to soil formation. Indeed, the mineral fraction coming from the parent rock by combining with organic molecules released by plants, alive or dead, leads to formation of the organomineral complex which characterises the soil. Rhizodeposition consists of different organic compounds, some released actively (secretions, mucilages) and others passively (exudates, lysates, mucigel) (Fig. 7.1).

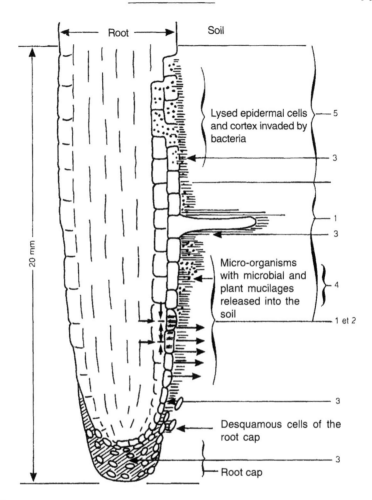

Root   Soil

Lysed epidermal cells — 5
and cortex invaded by
bacteria
— 3

— 1
— 3

Micro-organisms
with microbial and
plant mucilages   4
released into the
soil
— 1 et 2

— 3

Desquamous cells of the
root cap
— 3

Root cap

20 mm

Fig 7.1   Sketch of a root showing the origin of various organic compounds in
the rhizosphere: 1—Root exudates; 2—Secretions; 3—Mucilages; 4—
Mucigel; 5 — Lysates (after Rovira et al., 1979).

Exudates are small molecules, soluble in water or volatile, released
passively by living cells. These molecules are of various types:
carbohydrates, organic acids, amino acids, fatty acids, sterols, vitamins,
enzymes, nucleotides etc. (Table 7.1). Even though there is some analogy
between the exudates produced by many plants, differences do exist
between plant species and within a given species according to the stage of
plant development.

Table 7.1 Substances present in root exudates (after Curl and Truelove, 1986)

Classes of Compound	Nature of compounds studied	Most studied plant species
Sugars	Glucose, fructose, saccharose, maltose, galactose, rhamnose, ribose, xylose, arabinose, raffinose, oligosaccharides	*Triticum aestivum, Hordeum vulgare, Phaseolus vulgaris, Pinus* spp.
Amino acids	Asparagine, alanine, glutamine, aspartic acid, leucine, isoleucine, aminobutyric acid, glycine, cystine, cysteine, methionine, phenylalanine, tyrosine, threonine, lysine, proline, tryptophan, arginine, homoserine, cystathionine	*T. aestivum, Zea mays, Avena sativa, Pisum sativum, Phalaris* spp., *Trifolium* spp. *Oryza sativa, Gossypium barbadense, Lycopersicon esculentum, Pinus* spp., *Robinia pseudo-acacia, Bouteloua gracilis*
Organic acids	Tartaric, oxalic, citric, malic, acetic, propionic, butyric, succinic, fumaric, glycolic, valeric, malonic	*T. aestivum, Z. mays, P. vulgaris, L. esculentum, Brassica* spp., *Pinus* spp., *R. pseudo-acacia, P. vulgaris, Arachis hypogaea*
Fatty acids and sterols	Palmitic, stearic, oleic, linoleic, linolenic, cholesterol, stigmasterol, sitosterol	*P. vulgaris, A. hypogaea*
Growth factors	Biotine, thiamine, nicocine, panthotenate, choline, inositol, pyridoxine, amino-benzoic acid, n-methyl nicotinic acid	*T. aestivum, Phalaris* spp., *P. vulgaris, Pisum sativum, Trifolium* spp., *Medicago* spp., *Gossypium barbadense*
Nucleotides, flavonoids and enzymes	Flovonone, adenine, guanine, uridine, cytidine, phosphatase, invertase, amylase, protease, polygalacturonase	*T. aestivum, Z. mays. P. sativum, Trifolium* spp.
Various compounds	Auxines, scopolotine, fluorescent substances, hydrocyanic acid, glycosides, saponine (glucosides), organic phosphate compounds, factors of nematode encystment substances attracting nematodes, stimulating or inhibiting fungal mycelial growth, germination of spores and sclerotia, bacterial growth, stimulating germination of weed seeds	*Avena sativa, Medicago* spp., *Trifolium* spp., *P. sativum, L. esculentum, Lactuca* spp., *Fragaria vesca, Musa paradisiaca, Z. mays*

Secretions are compounds actively released by plants into the external evironment.

Lysates constitute the cell content released after autolysis of old epidermal cells of the root wall.

Mucilages are made of polysaccharides, amino acids and proteins. These compounds have diverse origins. They may be due to (i) secretions from vesicles of the Golgi apparatus in the cells of the root cap; (ii) hydrolysates of polysaccharides from the walls of cells located between the epidermis and the desquamous cells of the root cap; (iii) secretions by cells of epidermis lacking a primary wall; and (iv) bacterial degradation of the primary wall of dead cells of the epidermis.

Mucigel corresponds to gelatinous compounds of polysaccharide nature produced both by roots and microbial populations in the rhizosphere. This gel contributes to the contact between the soil particles and the root surface (Photo 7.1) and thereby improves the transfer of mineral elements and water towards the root. Mucigel also provides a lubrication enabling a root to penetrate the soil.

Besides the release of organic compounds, the rhizosphere is the centre of gas and ion exchanges. The partial pressure of O_2 and CO_2 is mainly governed by respiration of the roots and micro-organisms. Rhizospheric micro-organisms are adapted to the low concentrations of O_2 in the rhizosphere. Some microaerophilic populations show optimal growth and activity at low concentrations of O_2. Part of the microbial populations may use nitrate (NO_3^-) as an electron acceptor instead of O_2. This adaptation may also originate in the host plant. As an example,

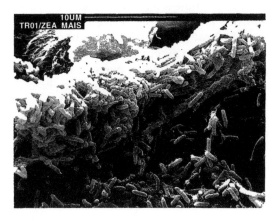

Photo 7.1 Scanning electron microscope picture of the colonisation of a maize root by a strain of *Rahnella aquatilis* (Achouak, Villemin, LEMIR, CPB-CNRS, Vandouevre les Nancy).

roots of rice plants have intercellular spaces (aerenchyma) in which oxygen coming from the leaves is accumulated and released into the rhizosphere, thereby reducing the oxygen deficit created by water. With this supply of oxygen, most bacterial populations in the rice rhizosphere are microaerophilic and not strictly anaerobic.

The balance between different ions in rhizospheric soil differs from that in non-rhizospheric soil. This difference is associated particularly with selective absorption of specific ions and release of others by the plant. This double exchange contributes to a change in pH of rhizospheric soil. So the pH of rhizospheric soil may vary from 1 to 2 units compared to that of non-rhizospheric soil. These variations are due essentially to the emission of HCO_3^- and H^+ ions. This phenomenon is influenced by the uptake of nitrogen by the plant. The absorption of nitrates by the plant triggers its release of HCO_3^- and OH. In contrast, absorption of nitrogen in the ammoniacal form by the plant causes the release of protons and acidification of the rhizosphere. Apart from these variations in pH, rhizospheric soil is characterised by a lower concentration of phosphorus and potassium than in non-rhizospheric soil. This difference is explained by the rapid absorption of these elements and by their limited diffusion in the soil. On the other hand, calcium, a mobile ion, tends to accumulate in the rhizosphere.

Rhizospheric Effect: A Temporal and Spatial Dynamic Concept

Localisation of the rhizospheric effect evolves in space according to root growth. Root exudation occurs mainly in the apical or subapical part of roots. This zone, which determines the rhizospheric effect, moves in the soil at the rate of root growth. This movement is especially well illustrated in the case of wheat (Fig. 7.2). Initially, the adhering soil clings to all of the seminal root system except the root tips. Then this sheath of adhering soil disappears from the oldest parts of the root and remains only on the youngest parts above the root tips. The rhizospheric effect also changes over time depending on the photosynthetic activity of the plant. Rhizodeposition corresponds to about 10 to 20% of the carbon of the photosynthetates. Reduction in photosynthetic activity, especially during ripening of wheat grains, is associated with a reduction in rhizodeposition and therefore in the rhizospheric effect.

The rhizosphere is very special environment wherein the balance of carbohydrates and mineral elements differs significantly from that in the non-rhizospheric soil. This difference between rhizospheric and non-rhizospheric soil is responsible for the high microbial density and activity found in the rhizosphere.

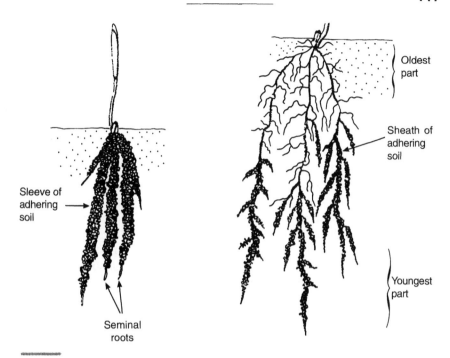

Fig 7 2 Displacement of soil adhering to the roots of wheat (*Triticum sativum* L.) over time (after Van Tieghem, 1886).

A PLACE OF INTENSE MICROBIAL ACTIVITY

Soil is an oligotrophic environment. Most soil-borne micro-organisms are heterotrophic for carbon and therefore dormant until a source of carbohydrates enables their activity. Rhizodeposition from roots constitutes an important supply of carbohydrates. Hence microflora is stimulated in the rhizosphere of the plant.

This stimulation leads to an increased microbial density in rhizospheric soil compared to bulk soil. It is expressed by the ratio of microbial density in the rhizosphere (R) and the soil (S). The different groups of micro-organisms may be classified in descending order of their R : S ratios as follows: bacteria, actinomycetes, fungi, protozoa and algae, microfauna. The R : S ratio of bacteria, for example, ranges between 10 and 20 but may reach values higher than 1000. The most favoured bacterial groups in the rhizosphere are mobile and hence likely to respond by chemotaxy to the root exudates, and have a high growth rate. Thus pseudomonads and enterobacteria, characterised by their mobility and high multiplication rate, are particularly numerous in the rhizosphere. The growth of fungi is also stimulated, but to a lesser extent, by root exudates. As in the case of bacteria, these exudates exert a chemotactic effect on mobile fungal spores (zoospores). They also help germination and elongation of the germ tubes, which results in fungal infection of the root. Even when the density of fungi in the rhizosphere is lower than that of bacteria, considering the large size of fungi, the fungal biomass is probably at least as much as that of the bacteria.

The rhizosphere causes not only a quantitative but also a qualitative modification in microbial populations. As examples, it was shown that diversity in populations of *Paenibacillus polymyxa* and fluorescent pseudomonads is reduced in the rhizosphere. This reduction in diversity is associated with selection by the host plant of micro-organisms that have enzymatic systems allowing metabolism of carbohydrates or specific amino acids specifically present in the root exudates. Such micro-organisms have a decisive competitive advantage compared to other micro-organisms. This advantage has been clearly demonstrated for the species *Agrobacterium tumefaciens*. This bacterium, responsible for the formation of crown gall on the host plant, diverts a part of the plant metabolism for its own benefit. Indeed, this bacterium introduces into the plant genome plasmid DNA that encodes synthesis of a group of amino acids called opines which are used exclusively by *Agrobacterium tumefaciens* as a source of carbon, energy and nitrogen. Selection by the host plant of particular microbial populations in also based on the ability of the micro-organisms to degrade certain toxic substances produced by the plant, for example phenolic compounds. Lastly, an important component of the selection achieved by the host plant is based on the specific recognition between micro-organisms and the plant. This mechanism was demonstrated when studying symbiosis between leguminous plants and nitrogen-fixing bacteria (*Rhizobium, Mesorhizobium, Bradyrhizobium*). Some compounds (flavonoids) present in the root exudates of leguminous plants attract strains of *Rhizobium sensu lato* and activate the expression of particular genes in these bacteria. These genes encode synthesis of recently characterised molecules which are involved in the recognition of infection in a plant. A specific recognition between bacteria and plants has also been described for non-symbiotic bacteria. A glycoprotein on the surface of bean root causes agglutination of specific isolates of pseudomonads. The agglutination of these bacteria allows their adhesion to the root and then root colonisation. Isolates showing the ability to agglutinate with the corresponding plant protein are dominant in the rhizosphere of the host plant. Specific recognition between host plant and micro-organism has also been described for some fungi (*Pythium, Phytophthora*). Adhesion of the zoospores of *Phytophthora cinnamomi* could be associated with the presence of specific carbohydrates in the mucigel surrounding maize roots.

The rhizospheric effect is characterised by a stimulation of microbial metabolism. Bacteria present in the rhizosphere appear to be larger than those in the soil during electron microscopic observations. Similarly, the germination rate of fungal spores is greater in the rhizosphere than in the soil at large. Recent progress in molecular biology allows one to confirm the great acitivity of rhizospheric micro-organisms and to specify the effects of the environment on the *in situ* synthesis of some metabolites. The principle of the method used consists of fusing, so-called reporter genes with the operon that encodes the activity studied. These genes synthesise compounds that are easy to visualise and/or quantify. An example of reporter genes are those whose activity is noted by the emission of light. These genes have cloned from marine bacteria that emit light naturally. The emission of light is due to the activity of an enzyme known as luciferase, produced only when the operon to which the regulator genes have been fused is active. Production of light indicates production *in situ* of the metabolite studied. Microbial metabolism is influenced not only by the quantity of root exudates, but also by their quality. Synthesis of some bacterial metabolites therefore requires the presence of particular substances. For example, synthesis of hydrocyanic acid by some fluorescent pseudomonads is possible only in the presence of glycine. Similarly, synthesis of indoleacetic acid requires the presence of L-tryptophan. The composition of root exudates largely determines the nature of rhizospheric microbial activities. These activities are also influenced by the particular physicochemical environment of the rhizosphere. Iron (Fe^{3+}) avilability is reduced because of plant and microbe uptake of this ion, associated with the low partial pressure of oxygen, leading to increased microbial synthesis of siderophores. Siderophores (etymology: *sideros*: iron, *phore*: bearer) are molecules of low molecular weight, soluble in water, produced by most micro-organisms when there is an iron shortage. They have a high level of affinity for iron (Fe^{3+}) and enable micro-organisms to acquire the iron necessary for their metabolism.

Localisation of Micro-organisms Varies in the Rhizosphere

Micro-organisms may be associated more or less intimately with the root. Some are present in the rhizospheric soil surrounding the root, which is influenced by root exudates. Others may adhere to the root surface (rhizoplane). Ectophytic micro-organisms are not evenly distributed along the root. Their density is higher in zones corresponding to the release of a large amount of carbon compounds such as the sites of formation of secondary roots. Some micro-organisms are endophytes. Among them, the most studied because of their agronomic importance are the nitrogen-fixing symbiotic bacteria (*Rhizobium sensu lato*) and mycorrhizal fungi. Infection of the root hairs of leguminous plants by *Rhizobia* leads to the formation of nodules in which atmospheric nitrogen is reduced. Formation of these nodules follows several steps: chemotaxy, reciprocal specific recognition between the bacteria and plant (previously described), curling of the root hair due to the effect of indoleacetic acid produced by the bacteria, production of an enzyme (polygalacturonase) that enables penetration of bacteria through the infection thread into the plant cells, multiplication of plant cells and formation of a nodule, modification of the morphology and physiology of bacterial cells (bacteroids) (Photo 7.2). The mycorrhizae (literally fungal roots) correspond to the association of a root and a mycorrhizal fungus. There are two main types of mycorrhizal association: ectomycorrhizal and endomycorrhizal. In both cases the fungus infects the root but its development *in planta* differs subsequently. Ectomycorrhizal fungi (basidiomycetes, ascomycetes or zygomycetes) form a parenchymatous

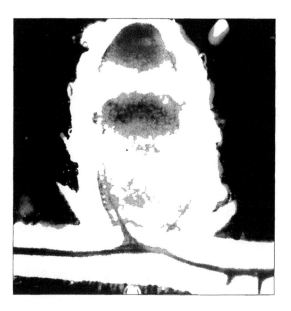

Photo 7.2 Optical microscopic observation (× 50) of a nitrogen-fixing nodule on a root of lucerne (*Medicago sativa*). An apical meristem ensures growth of this type of nodule. The vascular traces and endoderm of the nodule are peripheral in relation to the central fixative zone (Truchet, Camut, de Billy, Odorico and Vasse, 1989, INRA, Toulouse).

envelope of more than 40 μm that may constitute 40% dry mass of the mycorrhizae. Ectomycorrhizae are common in woody plants, both gymnosperms and angiosperms, and especially oak, beech, birch and conifers. In contrast to ectomycorrhizal fungi whose development remains cellular (Photo 7.3), that of endomycorrhizal fungi is both inter- and intracellular (Photo 7.4). Most of these fungi actually invade plant cells and form highly ramified mycelia, which explains why they are

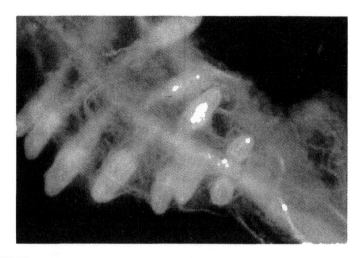

Photo 7.3 Optical microscopic observation (X 30) of an ectomycorrhiza of *Picea abies* (Lapeyrie, INRA, Nancy).

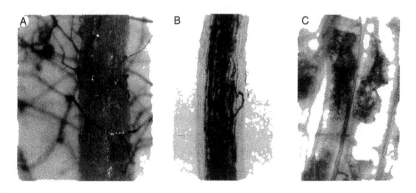

Photo 7.4 Optical microscopic observation of on arbuscular mycorrhiza in parsley (*Petroselinum crispeum*). Development of the external rhizomycelium infecting the root at the level of the appressoria (A. 1 cm = 0.2 μm), development of the fungus in the internal cortex of the root (B. 1 cm = 0.2 μm), detail of the fungus at stage B showing the intercellular hyphae and the intracellular arbuscles (C. 1 cm = 20 μm) (Gianinazzi-Pearson and Gollotte, CMSE, INRA, Dijon).

called arbuscular mycorrhizal fungi (zygomycetes). These types of fungi establish in association with almost all plant species of agricultural importance: wheat, maize, potato, soy bean, tomato, tulip, apple etc. Other micro-organisms are also endophytes. These are plant pathogenic bacteria (*Agrobacterium tumefaciens, Ralstonia solanacearum*) and fungi (*Fusarium oxysporum, Rhizoctonia solani, ...*) or non-pathogenic micro-organisms (*Pseudomonas* spp., *Fusarium oxysporum...*).

Rhizosphere: The Centre of Complex and Varied Microbial Interactions

These interactions may be classified schematically into 5 types: competition, antibiosis, parasitism, predation and synergy. The production of root exudates stimulates the development of rhizospheric microflora. However, this microflora soon reaches such a density that the quantity of available nutrients becomes limiting for carrying microbial growth and activity. Micro-organisms then enter into **competition** not only for nutrient sources but also for space, especially when their distribution along the root is irregular. Competition is even more intense when they have close nutrient and ecological requirements. As an example, the competition between pathogenic and non-pathogenic strains of *Fusarium oxysporum* is especially severe. It is partly responsible for the natural suppressiveness of some soils to fusarium wilts, major soil-borne diseases caused by pathogenic *F. oxysporum*. The substrates that are the object of competition between micro-organisms are essential for microbial metabolism and are present in limited concentrations. This applies to carbohydrates. There is also competition for some ions such as iron (Fe^{3+}). Most micro-organisms benefit from a strategy of acquisition based on the synthesis of siderophores released from the cell. Siderophores chelate iron (Fe^{3+}). These iorn chelates are recognised more or less specifically by receptor proteins localised in the microbial membrane. Depending on the micro-organisms, this strategy is more or less efficient. As an example, pseudomonads produce siderophores that have an affinity for iron (Fe^{3+}) which is higher than that of siderophores produced by fungi. Pseudomonads can thereby reduce the saprophytic growth of fungi in the rhizosphere by reducing the availability of iron.

Many micro-organisms (bacteria and fungi) synthesise a large variety of **antibiotics**. In spite of their degradation and adsorption on soil colloids, antibiotics are present in the rhizosphere at concentrations high enough to reduce the growth of susceptible micro-organisms. Only micro-organisms tolerant or resistant to these antibiotics develop in the rhizosphere.

Some fungi and bacteria produce **lytic enzymes** that contribute to fungal parasitism. The high bacterial density in the rhizosphere is associated with the presence of protozoa which develop at the expense of bacteria (predation).

Lastly, as opposed to the earlier described negative interactions, there are examples of synergy between micro-organisms. Thus some bacteria encourage infection of the roots of a host plant by the ectomycorrhizal fungus *Laccaria laccata*.

All microbial interactions, positive (synergy) or negative (competition, antibiosis, lysis, predation), lead to a precarious balance that is disturbed according to the quantitative and qualitative variations in rhizodeposition by the plant.

RHIZOSPHERIC MICROFLORA INFLUENCES GROWTH AND HEALTH OF THE HOST PLANT

A great number of soil-borne micro-organisms, stimulated by root exudates, develop in the rhizosphere and express their pathogenicity on the plant. These pathogenic agents may or may not be parasites. Non-parasitic pathogenic micro-organisms decrease growth of the host plant by releasing toxic metabolites. Decreases in yields recorded in the Netherlands in potato fields cultivated continuously on the same plot, have been ascribed to the development of non-parasitic fluorescent pseudomonads—producers of hydrocyanic acid. These bacteria are called deleterious. Among pathogenic agents, some are strict parasites. This applies to the *Plasmodiophora brassicae* fungus, which is responsible for clubroot of crucifers, preserved in the soil in the form of oospores. These spores are in fungistasis until a root of a host plant comes in contact with them, stimulates their germination by the production of exudates, and thereby enables infection of the root by the fungus. The development cycle of most pathogens occurs in two phases: first as a saprophyte, then as a parasite. Pathogenic *F. oxysporum* have a more or less active saprophytic stage depending on availability of carbon sources. This saprophytic phase is interrupted when the fungal mycelium infects a root and then colonises the vascular system of the host plant (parasitic phase). After the death of the parasitised plant, due to obstruction of the vessels, the pathogenic agent behaves again as a saprophyte.

The health of plants depends not only on the density and activity of pathogenic agents, but also on the density and activity of antagonistic micro-organisms present in the rhizosphere. Depending on the relative extent of these antagonists, a plant will be more or less healthy. In fact, the severity of a disease of telluric origin for plants belonging to the same variety and infested with the same pathogenic inoculum, varies with the extent and activity of the rhizospheric microbial biomass. The higher and more active the biomass, the less the saprophytic growth of the pathogenic agent (bacteriostasis and fungistasis) and consequently the likelihood of a contact between the pathogenic agent and the host plant is lower. This mechanism is called a general suppressiveness. Against the background of general suppressiveness the mechanisms of specific resistance associated with the antagonistic activity of particular microbial populations can be expressed. Their antagonistic activity is due to the production of metabolites toxic to the pathogen. As already described, these are mainly siderophores, antibiotics, enzymes as well as bacteriocins. The two mechanisms, general suppressiveness and specific suppressiveness, are expressed with particularly high intensity in the rhizosphere of soils called suppressive. In these soils, in spite of the presence of the pathogenic agent, disease severity remains very low. A well-studied case is the natural suppressiveness of some soils to fusarium wilts. This suppressiveness has been ascribed to all the microflora (general suppressiveness) and to two microbial groups, non-pathogenic *Fusarium oxysporum* and fluorescent pseudomonads (specific suppressiveness). Apart from a reduction in saprophytic growth of the pathogenic agents (microbial antagonism), some beneficial micro-organisms may also reduce their parasitic growth by simulating the natural defence reactions of the host plant. Induction of resistance in the plant against fusarium wilts achieved by a strain of fluorescent *pseudomonad* has been ascribed to the presence of particular lipopolysaccharides in its membrane.

Modification of the root environment, microbial balances and of root physiology by mycorrhizal fungi is also followed by an improvement in the health of the host plant, which is less susceptible to soil-borne diseases.

Certain rhizospheric micro-organisms also exert a beneficial effect on plant growth. Thus, some improve the mineral supply to plants. These are mainly symbiotic micro-organisms. Accordingly, rhizobial bacteria fix atmospheric nitrogen for the benefit of the nodulated plant. This plant in turn provides the bacteria with the carbohydrates and amino acids necessary for their metabolism. Symbiotic fixation of atmospheric nitrogen is considerable; it represents approximately 160 Mt of nitrogen per year, which is twice the annual production of nitrogenous fertilisers. Similarly, mycorrhizal fungi improve the supply of phosphorus, nitrogen, and often of water to mycorrhized plants. Some free micro-organisms may improve the supply of nitrogen to plants (*Azospirillum, Paenibacillus, Burkholderia,* ....) and also of phosphorus (*Pseudomonas, Bacillus,* ...).

CONCLUSION

The rhizosphere is a centre of considerable microbial activity. This activity influences the growth and health of plants. Indeed, some micro-organisms are phytopathogenic, others improve the water and mineral supply to plants; these micro-organisms are subjected to the antagonistic activity of all the soil microflora and specific microbial population.

A better knowledge of these complex interactions now enables the agronomist and the plant pathologist to consider altering some microbial balances to benefit the plant, in order to reduce the use of inputs (fertilisers and synthetic pesticides). A change in these microbial balances may be achieved either by microbial inoculations or by favouring local indigenous beneficial microbrial populations.

Inoculation of soy bean with *Bradyrhizobium japonicum* precludes the use of nitrogenous fertiliser. Inoculation of cereals with *Azospirillum lipoferum* enables more effective use of nitrogenous fertiliser. Inoculation of woody plants with ectomycorrhizal fungi and both woody and non-woody plants with selected ecto- and endomycorrhizal fungi alternatively improves plant growth in soils in which the use of fertilisers is reduced and rational. Lastly, introduction of antagonistic micro-organisms such as *Fusarium oxysporum*, fluorescent pseudomonads, *Talaromyces flavus* etc. enables control of various soil-borne diseases.

Among environment friendly cultural practices that allow one to modify microbial balances in favour of the plant, the best known is crop rotation which, when carried out logically, helps disturb the development cycle of certain pathogens and thereby reduce the infection potential of soils. Modifications of the physicochemical properties of rhizospheric soil by certain agricultural practices also affects the severity of some soil-borne diseases. Raising the pH, after liming the soil, is followed by a reduced severity of fusarium wilt while lowering in pH after an ammoniacal nitrogenous fertilisation is followed by an increase in disease severity. Utilisation of transgenic plants which produce compounds favouring specifically indigenous beneficial micro-organisms is under investigation.

Management of plant-micro-organism interactions within the rhizosphere should help reduce inputs and thereby prompt practice of an agriculture more environment friendly.

Glossary

Ectophytic micro-organisms	Micro-organisms that remain and develop externally on a plant.
Endophytic micro-organisms	Micro-organisms present within a plant at the intra- and extracellular level. These micro-organisms may form or cause formation of particular functional structures.
Heterotrophic micro-organisms	Organisms that require the presence of organic compounds for their growth and multiplication. These organic compounds are used as sources of carbon and energy.
Oligotrophic environment	An environment poor in nutrients.

FURTHER READING

Crawford NM, Kahn ML, Leustek T, Long SR. 2000. Nitrogen and sulfur. In: Biochemistry and Molecular Biology of Plants. Buchanan BB, Gurissem K, Jones RL (eds.). American Society of Plant Physiologists, Rockville, MD, pp. 786-849.

Curl EA, Truelove B. 1986. The Rhizosphere. Advanced Series in Agricultural Sciences, no. 15. Springer-Verlag, Berlin, Heidelberg, NY, Tokyo, 288 pp.

Foster RC, Rovira AD, Cook TW. 1983. Ultrastructure of the Root-Soil Interface. The American Phytopathological Society, St. Paul, MN.

Lemanceau P. 1992. Effets bénéfiques de rhizobactéries sur les plantes: exemple des *Pseudomonas* spp. *fluorescens*. Agronomie 12: 413-437.

Lynch JM. 1990. *The Rhizosphere*. J. Wiley & Sons, Chichester, NY, Brisbane, Toronto, Singapore, 458 pp.

Pinton R, Varanini Z, Nannipieri P. 2001. The Rhizosphere Biochemistry and Organic Substances at the Soil-Plant Interface. Marcel-Dekker, Inc.

Read DJ, Lewis DH, Fitter AH, Alexandre IJ. 1992. Mycorrhizes in Ecosystems. CAB Int'l, Univ. Press, Cambridge, 419 pp.

Waisel Y, Eshel A, Kafkafi U. 1996. Plant Roots. The Hidden Half. Marcel-Dekker, Inc. (2nd ed., revised and expanded).

Part III

Consequences of Exploitation of Soils by Man

Part Outline

CHAPTER 8

Salinisation of Soils

C. Cheverry, G. Bourrié

THE RISKS

A Global Problem

Salinisation of soils has already reduced, or will reduce in the short term, a substantial part of cultivated areas in the world. This phenomenon is due to an excessive accumulation of highly soluble salts (chlorides, sulphates and carbonates of sodium or magnesium) in the surface horizon of soils, which is observed by a reduction in soil fertility (Fig. 8.1). The supply of water to plants is rendered more difficult; some elements may also have a specific toxic effect (Na, Cl, B, Se); sodium may be fixed on clays and concomitantly modify their behaviour in the presence of water. The overall physical properties of soil (permeability) are then degraded.

Salinisation may be 'primary', that is, inherited from natural conditions, for example the presence of saline geological layers. It can also be 'secondary', i.e., associated with human activity and in particular irrigation practices. It may also be considered 'potential' unless man modifies his present practices.

Fig 8.1 (Top): Saline soil with a surface accumulation of sodium chloride (Cheliff-Algeria plains) (photo by C. Cheverry). (Bottom): Saline soil with a surface accumulation of sodium sulphate (Mexico) (photo by J.-Y. Loyer).

It is estimated that 6.5%, or 9 million km^2, of the land in the world is already affected by this phenomenon, rising to 39% in arid zones (Qiguo, 1994). Soil salinisation mainly affects arid regions. Evapotranspiration in these regions is actually much larger than the precipitation (rainfall) during a large part of the year; soil water therefore rises due to capillary action and dissolved salts crystallise at the surface.

However, salinisation is observed in all continents and in all climatic regions of the world, including the polar circle (Szabolcs, 1994), and at all elevations, whether below sea level (region of the Dead Sea) or at 5000 m (Tibetan plateau) (Fig. 8.2). Even a country with a temperate climate such as France is affected. Camargue and the marshes in western France (Charente, Vendée and Poitou, totalling 300,000 hectares) have been subjected to influence of the sea salt. The risks of degradation of soils associated with residual salinisation here are real.

It must be emphasised that the problem of salinisation is even more serious since not only cultivated plants are threatened. Micro-organisms,

Fig 8 2 Soils affected by salts and their distribution (after Szabolcs, 1994).

invertebrates, vertebrates and man himself may react to an environment of soils and water that are too saline, and this adaptation may result in a profound change in the food chain (Miyamoto, 1994).

Acute Problem Which Has Recently Increased

The risk of losing new areas of land due to salinisation has greatly increased over the last 20 years because of a considerable increase in irrigation. Since 1949, irrigated areas have increased by a factor of four and now cover more than 270 Mha (Table 8.1). As a result of poor water management in these areas, their greater part is threatened by rapid salinisation. It is estimated that 10 Mha of irrigated land are lost for cultivation every year due to salinisation and that half of all irrigated areas are threatened in the long term. The phenomenon may also be a rapid one: in Argentina, 50% of the 40,000 ha irrigated in the nineteenth century are now too saline for use. In many countries, such degradation has taken place in just a few decades: Pakistan (Indus Valley), Peru, Australia, Iraq, Syria,....

Competition for water has recently become more acute. In spite of the construction of large dams in many countries providing considerable water reserves, there is not enough water for irrigation. Farmers are thus forced to use waters of poorer quality for irrigation (recycled irrigation water, waste water from towns,...). The risk of secondary salinisation of soils has thereby increased.

In highly developed countries, the risk of salinisation has now acquired a new perspective. The quality of drainage water taken from

Table 8.1   *Increase in irrigated area throughout the world since 1800 (after Szabolcs, 1994)*

Year	Irrigated area (Mha)
1800	8
1900	48
1949	92
1959	149
1980	230
1990	265

irrigated areas and stored in water reservoirs subjected to evaporation is now questionable. Waters from an agricultural activity, even so traditional as irrigation, are beginning to attract public concern because of their salinisation and the large number of transitional elements or heavy metals they contain. They are considered as 'waste' waters, potentially hazardous for the environment, as are waste waters from industrial or urban activities, or even some effluents from livestock.

The most spectacular example is that of San Joaquim Valley in California, USA. Recent incidents (bird mortalities) ascribed to an excessively high selenium concentration in the water of the Kesterson Reservoir (which collects drainage water from irrigated areas) have prompted the state of California for ecological reasons to amend the law with respect to management of this type of water in that region (Miyamoto, 1994).

In general, the acute realisation that politicians now have that irrigation degrades the water quality for subsequent users has led to serious interstate negotiations in the USA, and between the USA and its neighbours—Canada and Mexico— with respect to the Colorado River. In the years to come, difficult negotiations may be expected between Turkey, Syria and Iraq, between India and Pakistan,.... The phenomenon of soil salinisation has unquestionably acquired a geopolitical dimension.

Problem Difficult to Treat Technically

One would think that human beings are well equipped to deal with this threat. Man has long been aware of the consequences of salinisation; abandonment of agriculture in certain areas (valleys of the Tigris and Euphrates rivers: Mesopotamian civilisation) has been largely attributed to this phenomenon. But this is only partially true. Some populations possess an ancient skill in managing irrigation systems and avoiding risk of salinisation (system of irrigation tunnels in southern Tunisia...). But very recent and sudden increase in irrigation has been observed in regions where users of soil and water have no experience. There can be no transfer of experience from one people to another, or even of technology, without adequate adaptation.

The case of China is significant. This country has become the world's main producer of cereals, due in particular to considerable developments in irrigation. Its farmers are highly skilled in agricultural practices, protecting the soil biology (good organic manure, ...). About 40 Mha in this country are affected by salts. Although in some irrigated areas, due to a voluntary management policy (improvement in drainage, lowering the water table level, biological measures), the Chinese have been able to check salinisation and restore fertility of land (this applies to the Huang Huai Hai plain), in other regions (Matsumoto et al., 1994), new areas are still being lost every year, for example in the Huang Ho delta.

Even in the most developed countries with high levels of technology and considerably experienced farmers (USA, Australia, Canada, Spain, ...) salinisation remains a constant threat. These developed countries certainly have efficient and even sophisticated tools (see Chap. 3) but experience has shown that they 'do not work in all situations'. Efforts to acquire better knowledge of the mechanisms of salinisation and research to find countermeasures adapted to different technical and social conditions are urgently needed.

SOURCES, MECHANISMS AND CONSEQUENCES OF SALINISATION

Principal Sources of Salts that Contaminate Soils

Four main sources of salts that are likely to accumulate in soils can be identified. The first is the **ocean**, which may carry through the **atmosphere** dissolved salts in rain-water and, in particular, very tiny hygroscopic particles in the form of aerosols. These salts fall to the soil mainly during rains but are also deposited in dry weather. This is clearly observable near seacoasts but may also be seen deep inside the continents to some extent (Bresler et al., 1982; Meybeck, 1984, 1987). Most of chloride and sulphate ions, a small but not negligible part of sodium and magnesium therefore come indirectly from the ocean. Under these conditions the hydrochemical facies of the waters is 'sodium-chloride or sodium-sulphate'.

The second source of salts is the **lithosphere** through weathering of rocks that form oceanic islands or the continental crust. In the first case, these are basalts and in the second, granites or similar rocks, and the dissolved salts result from reactions of hydrolysis of aluminosilicates and ferromagnesian minerals (feldspars, micas, amphiboles, pyroxenes or olivines). These processes use up the protons provided mainly by the dissociation of CO_2 in the soil atmosphere, which in turn comes from the oxidation of organic matter. During these reactions the most insoluble elements, Al, Fe and part of silica reprecipitate, resulting in the components of saprolites and soils (clays, iron oxides and aluminium hydroxides) while alkalis and alkaline earths migrate in solution in ionic form. The corresponding positive charges in solution are balanced by hydrogenocarbonate or carbonate ions. Further, the quantity of dissolved salts depends on the quantity of protons available and hence on the partial pressure of CO_2, rate of weathering of minerals, temperature, extent of division of minerals, and reactivity at the mineral-solution interface. This weathering is therefore of a climatic type (Tardy, 1993).

Intensity of weathering is maximum when drainage removes dissolved products (short residence time of water), when turnover of organic matter is rapid (high productivity and rapid decomposition) and when rock minerals are easily weatherable (unstable minerals), as on young mountains under a tropical or humid equatorial climate. The composition of soil solutions and drainage waters thus results from the addition of atmospheric contributions and those from the weathering of rocks from which net uptake by biomass must be subtracted in the case of growing crops or forests (Bourrié and Lelong, 1994). In rocks of the continental lithosphere, Na and Ca are more abundant than Mg and K, and waters therefore belong to 'calcium-bicarbonate' or 'sodium-bicarbonate' facies. This explains why 'calcium-bicarbonate' facies predominates more and more over 'sodium-chloride' facies as the distance from the ocean increases.

The third source of salts is simply dissolution of **fossil salts** contained in some geological layers. This is the principal origin of the phenomenon of primary salinisation observed throughout the world. The salts may come from evaporites (rocks) or soil solutions trapped in sediments of marine origin. A single example is given here: in the United States, in the Arkansas River, mineralisation of waters largely increases after the river flows through saliferous Permian deposits. The salts thus mobilised may be either of continental (Caucasus, Manchuria, Mongolia, Chad, ...) or of marine origin (especially ancient deltas), sometimes as a result of artesian phenomena, with a rise to the surface of fossil saline waters from great depths (Central Asia, North Africa, ...).

The fourth source corresponds to **anthropic contributions**, either directly as from salts used as fertilisers (KCl, $(NH_4)_2SO_4$ ...) or indirectly from atmospheric fall-outs, some nitrogen or sulphur compounds falling back in saline form and, foremost, through the supply of irrigation water already slightly saline. It may be noted that human beings intervene by supplying salts directly and also by global or regional changes in atmospheric contributions, climatic and hydrologic regimes, and most particularly through water management and modification of land use (encroachment of forests etc.).

Excessive Accumulation of Salts in Soils: Phenomenon Frequently Associated with Poor Management of Irrigation and Salt Balances

Secondary salinisation often results from poor management of the water and salt balance at the level of a plot. A classic case is of irrigations using waters from catchment areas (behind large dams or directly from rivers or streams). Farmers supply water much in excess of the actual requirements of plants. As a result, some of this water percolates, which leads to a gradual rise in level of the water table.

In many regions (the Indus Valley in Pakistan and the valley of the Niger River in southern Sahara are examples), it has been noted that an aquifer initially at a depth of 50 metres had risen over time to less than 2 metres below the soil surface. Capillary rise becomes active and salts dissolved precipitate at the soil surface, in the very horizons colonised by roots of cultivated plants: ithas been estimated that the salt concentration in irrigation water is increased by 2 to 20 times in the root zone.

The general salt balance can be written as:

$$Sp + Si + Sr + Sd + Sf = Sdw + Sc + Sppt \qquad (1)$$

where: Sp is the quantity of salts in the precipitations (rains) over a particular area; Si the quantity of salts contained in irrigation waters; Sr the variation in quantity of soluble salts present in soil solution at the beginning and end of a particular period; Sd the quantity of salts dissolved by weathering of minerals, or released by desorption (may be negative if the phenomenon of sorption predominates); Sf the quantity of salts in the fertilisers supplied; Sdw the quantity of salts removed from the area by drainage waters; Sc the quantity of salts removed by crops; and $Sppt$ the quantity of salts precipitated in a sparingly soluble form $(CaCO_3 ...)$.

If the salt content in soils is too large for the plants to obtain a proper supply of water, excess salts present in the zone explored by the roots must be leached until the salt concentration in the soil solution in that zone is reduced under a given threshold value. Irrigation may then be continued while ensuring a balanced regime. In this management process, the concept of a leaching fraction (LF) plays a very important role:

$$LF = Dd/Di = (Di - ET)/Di \tag{2}$$

where: Di and Dd are the levels per unit area for a given time of irrigation and drainage waters respectively; and ET is the evapotranspiration. It can be seen in Figure 8.3 for different types of saline soils in the United States that the effect of an increase in 'leaching fraction' on desalinisation is much larger when the soil solution has a large initial salt content.

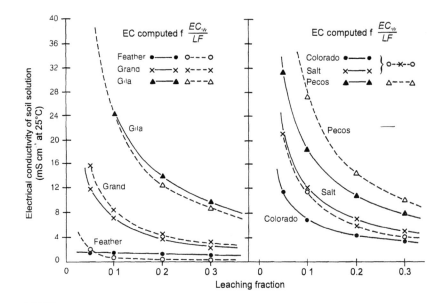

Fig 8.3 Computed electrical conductivity of drainage waters in a steady state, versus the leaching fraction. Solid lines: computations with the simulation model; dashed lines: computations from the electrical conductivity of irrigation water and the leaching fraction, on the assumption that there is neither precipitation nor dissolution of soil minerals (after Bresler et al., 1982).

Balances have been attempted at the level of large irrigated regions from hydrological data, e.g. in the Rio Grande catchment (USA) (Miyamoto et al., 1994). Water managers have therefore to attain simultaneously three goals: (i) to supply only the correct quantity of water to meet plant requirements; (ii) to maintain the salt concentration in the root zone a little below the threshold given (see next section); (iii) not too seriously affect the downstream users of drainage waters. Their task will be a difficult one and assumes proper assessment of the consequences that may follow errors in management.

Consequences of Salinisation

The first consequence is a result of a change in osmotic potential of the soil solution when the salt content increases. The electrical conductivity (EC) of this solution, measured using a resistance bridge or electrodes, is associated with the dissolved solid charge (DSC) by relations of the type:

$$DSC \ (mg \ l^{-1}) = 0.64 \ EC \ (mS \ cm^{-1}) \qquad (3)$$

(US Salinity Laboratory, 1954; mS = milli-Siemens)

The relation with the osmotic potential can be estimated from:

$$\Phi_0 \ (bar) = 0.36 \ EC \ (mS \ cm^{-1}) \qquad (4)$$

(US Salinity Laboratory, 1954)

Most plants are certainly capable of a certain regulation of their internal osmotic potential depending on that of the external medium, but this regulation, extremely variable depending on the plants (larger for halophytes), is limited (Katerji, 1995). The first consequence of salinisation is therefore the likelihood of plants being deprived of water. Reduction in yield based on salinity depends on the salt tolerance of the particular plants (Fig. 8.4).

The second consequence results from the specific role of certain ions which accumulate when the phenomenon of salinisation develops. For example, some fruit trees may be sensitive to an accumulation of sodium chloride. In other cases, an excessive accumulation of sodium indirectly causes shortage of calcium and magnesium (tomatoes, celery, ...). This

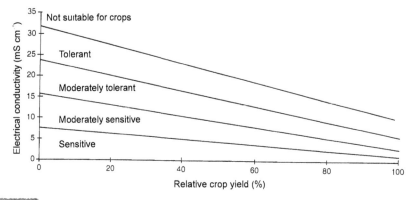

Fig 8.4   Salt tolerances of different crops (after Bresler et al., 1982).

has been the subject of many studies since Bernstein (1975). Recent research has stressed the accumulation of transitional elements (B, Se, As) or heavy metals (Cd, Hg,...) that may sometimes be associated to major ions (Na, Ca, Mg, HCO_3, Cl, SO_4) during salt accumulation. The

(a)

(b)

Fig 8 5 (a) Management of a mangrove soil (acid sulphate soil) in Casamance (Senegal); (b) Saline soil, sodium-carbonate, with black salt; polder of Lake Chad (Chad) (photo, C. Cheverry).

case of boron is often cited because in soil the difference between concentrations triggering phenomena of shortage in plants and those resulting in toxicity is narrow for many crops.

The third consequence results from changes in soil pH that may often follow the phenomenon of salinisation. Two cases are somewhat extraordinary. The first is a large rise in pH observed when sodium hydrogenocarbonates or carbonates accumulate (process of 'alkalinisation') (Valles et al., 1998). The soil pH may rise to levels of 9 to 10. The soil organic matter dissolves (Bertrand et al., 1994). The fertility of these soils is then considerably reduced because many elements essential for plants become totally insoluble at this pH.

An opposite case is that of acid sulphate soils in which the pH may fall to levels between 2 and 4. This occurs frequently in mangrove swamps (Fig. 8.5) managed with poorly controlled drainage because sulphides accumulated at a shallow depth in the sediments are oxidised, thereby releasing sulphuric acid. If the soil does not initially contain large stocks of $CaCO_3$, acidification inevitably follows and the soil is sterilised (aluminium toxicity, etc.). Table 8.2 lists the major chemical types of saline soils in the world and the effect of salinity on fertility.

Table 8.2 *Major categories of saline soils*

Types of soils affected by salts	Electrolytes causing salinity or alkalinity	Environment	Properties affecting the living medium	Counter-measures
Saline soils (Solonchaks)	Sodium chloride and sulphate	Arid or semi-arid	High osmotic pressure of soil solution, toxic effect of chlorides	Leaching of excess salts
Alkali soils (Solonetz)	Sodium ions	Semi-arid, semi-humid and humid	High pH level, poor physical conditions, calcium deficiency	Lowering or neutralisation of high pH levels by chemical amendments
Magnesian soils (Magnesic solonetz)	Magnesium ions	Semi-arid and semi-humid	Toxic effects, high osmotic pressure, calcium deficiency	Chemical amendments; leaching

(Table 8.2 Contd.)

(Table 8.2 Contd.)

Gypsisols	Calcium ions (mainly CaSO$_4$)	Semi-arid and arid	Low pH level, deficiency of some nutritive elements	Alkaline amendments
Acid-sulphate soils (Thionic Fluvisols)	Iron and aluminium ions (mainly sulphates)	Coastal areas and lagoons	Acidic level, toxic effect of aluminium, deficiency of some nutritive elements	Liming

Phenomenon of Sodisation: Key Role of Clays, Their Mineralogy and Organisation

Accumulation of salts on the surface is not the only risk for the fertility of soils due to salinisation. Fixation of sodium cation on clays (process of 'sodisation') is another aspect of the problem. When the quantity of fixed sodium is significant and the concentration of electrolytes in the medium is low, clays have a tendency to swell and disperse in the presence of water. The hydraulic conductivity of the soil is reduced (Fig. 8.6). The pores are constricted, soil surface becomes impermeable, crust formation occurs.... This may result in increased runoff or erosion.

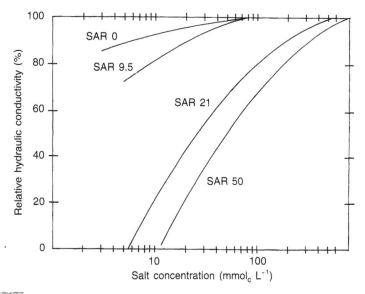

Fig. 8.6 Effect of the sodium adsorption ratio (SAR) and salt concentration on hydraulic conductivity of a soil at Lindley (Aridic Paleustalf; after Sumner, 1993).

This has often been observed not only in saline irrigated areas when the only objective, for reasons of economy, was to 'get rid' of the salt so as to reduce the osmotic potential without simultaneously attempting to eliminate the exchangeable sodium, but even in other regions (Central Africa, Australia,...), where salinisation occurs even though the soils are not highly saline. Only 5% sodium in the adsorption complex of soils can result in physical degradation.

This involves ion exchange on the negatively charged surface of clays (Rieu et al., 1991). Adequate knowledge of the laws regulating competition for the exchange sites is therefore necessary here. From this viewpoint the two important parameters for estimating the hazards of sodisation of irrigated soils are the SAR (sodium adsorption ratio) of the irrigation waters or the soil solution:

$$SAR = Na^+/(Ca^{2+} + Mg^{2+})^{1/2} \text{ (concentration of ions in mmol } L^{-1}) \qquad (5)$$

and the percentage of sodium fixed on the adsorbent complex, ESP

$$ESP = Na^+{}_{exch}/CEC \text{ (cation exchange capacity) (in meq } L^{-1}) \qquad (6)$$

An empirical relation that links the two values was used for a long time:

$$ESP = 1.475 \, SAR/(1 + 0.0147 \, SAR) \qquad (7)$$

The relation was later improved based on progress in geochemistry (considering the existence of ion pairs...). Formulations vary in regional context and are not given here. Figure 8.6 clearly shows that the reduction in hydraulic conductivity of a soil is even larger when the SAR is large and the concentration of salts low. The major points for a more detailed analysis of these phenomena are: firstly, a good knowledge of clay mineralogy. Figure 8.7 shows, in the case of South African soils, that soils with a high content of montmorillonite and micas are more sensitive to physical degradation than those with a high kaolinite and haematite content.

Many investigations have been conducted over several years on the role of the microscopic organisation of clays in a sodic environment and on the phenomenon of release of Ca and Na ions (Shainberg, 1968) and also on the corresponding degree of swelling and dispersion (Daoud and Robert, 1992; Cheverry and Robert, 1993). A detailed report was recently presented (Sumner, 1993). Introduction of the concept of 'dispersion potential' (Pdis) appears highly promising:

$$Pdis = Ptec - Psol, \qquad (8)$$

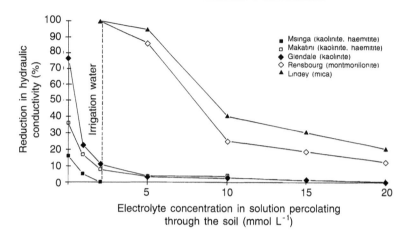

Fig 8 7   Relationship between reduction in hydraulic conductivity and reduction in electrolyte concentration at constant values of SAR for several irrigated soils in South Africa (after Sumner, 1993).

where Ptec is the osmotic pressure necessary to flocculate the particles and Psol the osmotic pressure of the liquid phase in soil. Figure 8.8 classifies soils according to their dispersion potential.

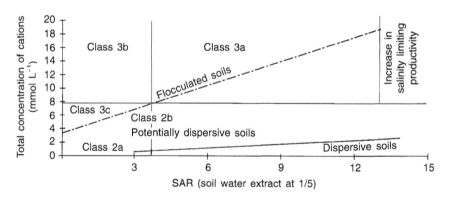

Fig 8 8   Classification of dispersive soils based on the relationship between SAR and total concentration of cations (TCC), both measured in an extract of soil at 1/5 (after Sumner, 1993).

The strategy adopted against sodisation is specific and often involves a supply of calcium in a sparingly soluble form (gypsum) to replace sodium on clays. Acidifying treatments may sometimes be adopted with this, or polymers may be used to stabilise the hydraulic properties of the affected soils.

SOME STRATEGIES TO BE ADOPTED AGAINST SO-CALLED SALINISATION (EXCESS OF SALTS)

Effective Tools for Follow-up of Soil Salinity

The salts under consideration are extremely mobile because of their high solubility. There is therefore considerable spatial and temporal variability in their concentrations and it is useful to take many measurements of the electrical conductivity of soils. This is now possible due to further developments in systems with batteries of electrodes connected in series with devices for electomagnetic induction. These are fitted on mobile tractors equipped with GPS (global positioning systems). The various measurements of apparent electrical conductivity obtained in this manner are treated by geostatistics or MLR (multilinear regressions). In California, these methods have been improved (Rhoades, 1994) and applied and adapted in many other countries, for example Tunisia (Job, 1992).

Good Knowledge of Geochemical Mechanisms

Salinisation, whether primary or secondary, develops as the concentration of solutions increases through evaporation, even from waters that initially had a low salt content: irrigation waters are often soft waters and have a calcium-bicarbonate facies rather than a sodium-sulphate one when they are taken from major rivers that originate in montane regions where hydrolytic weathering predominates. When evaporation is low or when the system is well drained and the leaching fractions well balanced, the dissolved salts return to rivers and groundwater and then to the sedimentation basins or the world ocean. But when evaporation is considerable, as in the floodplains of large rivers or especially in endoreic basins, salts precipitate in the soils. Salts are often deposited in the form of regular halos depending on the factor of concentration of water in the area (Cheverry, 1974; Montoroi, 1994). The most frequently observed salts are sodium sulphates (mirabilite and thenardite), sodium and magnesium sulphates (bloedite) or magnesium sulphates (epsomite) and sodium bicarbonate and carbonates (nahcolite, natron, trona).

In acid sulphate soils (Thionic Fluvisols), original parageneses are observed, with minerals of the jarosite and alunite group (Le Brusq et al., 1987; Montoroi, 1994). Since the solutions are highly concentrated, calculation of the degrees of saturation of minerals based on analysis of the waters, necessitates consideration not only of the long-range interactions between ions, but also the short-range interactions (Van der

Waals type) between the ions and the solvent (hydration spheres, disturbance of hydrogen bonds). This is usually done in two ways: i) using the laws on the coefficients of activity extending empirically the Debye-Hückel law, with provision for the formation of ion pairs (Al Droubi, 1976), or ii) by overall thermodynamic considerations of ionic interactions (Pitzer and Simonson, 1986). This is in the domain of physical chemistry of electrolyte solutions and is not discussed further here.

Models are now available based on any of these approaches which help in simulating the changes in composition of a solution during its concentration by evaporation and in calculating the nature and quantity of precipitated salts. Variations in different parameters such as pH, ionic strength, electrical conductivity, activities and concentrations of ions, the SAR etc., are given by these models depending on the concentration ratio of waters by evaporation.

The sequence of precipitation of salts therefore depends on the initial composition of soil solution, as influenced by inputs from the oceans and continents, and on the interaction with sedimentary material, mostly inherited, which forms the soils of these plains (Gac, 1979; Bourrié and Lelong, 1994; Vallès et al., 1995). This interaction consists of dissolutions and ion exchanges with clays. The sequence of salt precipitation starts generally with calcite, followed by gypsum (Al Droubi et al., 1976; al Droubi et al., 1980), generally combined (for example, in Vertisols) with aluminosilicates or magnesium silicates of the group of 2/1 clays (montmorillonite, stevensite) or the group of fibrous clays (attapulgite, sepiolite). This removes part of Ca, Mg and Si from the solutions, leaving a residual water enriched with Na and K. Depending on the proportions of hydrogenocarbonate and carbonate on the one hand—functions of alkalinity and pH (Bourrié, 1976)—and chlorides and sulphates on the other, the solution will move either towards the carbonate facies (alkaline route) or the chloride-sulphate facies—the neutral saline route. The concept of residual alkalinity helps to explain the dichotomy observed in the development of the system (Al Droubi, 1976; Al Droubi et al., 1980; Vallès et al., 1995) and especially variations in pH.

The salts that come from the sea salt through the atmosphere (NaCl and $MgSO_4$) are separated by geochemical processes in soils. NaCl is transported passively to the ocean or to endoreic salt lakes, while Mg reprecipitates mostly in the form of magnesium silicate, which 'consumes' alkalinity by reverse hydrolysis, releasing sulphate in exchange. Ca and Na released by hydrolytic weathering of continental rocks feed new formations of carbonates and sulphates (calcite, gypsum, mirabilite or thenardite, trona or natron). Ions exchange between clays and solutions may locally modify this evolution by changing the ratios between cations. This results in turn in modifications of the physical properties of soils, if the existing clays are able to swell (see above), and consequently the circulation of air and water. In the presence of organic matter and when the temperature is favourable for biological activity, the environment may become favourable for reduction of sulphates to sulphides, which leads to an alkaline milieu.

An extremely important case from this point of view is the continent-ocean interface. In the tropical zone where this interface is formed of plains that can be flooded by the sea (Florida, Senegal, Thailand,...), the growth of mangroves provides organic matter which enables considerable reduction of sulphates to sulphides deposited. In this process, the alkalinity produced is neutralised by sea-water. Later oxidation of the stock of sulphides is responsible for formation of acid sulphate soils (see above) (Van Breemen, 1976).

The salt-solution-soil systems are systems with rapid kinetics and the nature of the salts, including their degree of hydration, may change rapidly laterally, vertically or chronologically. They are also biological environments with severe constraints (osmotic potential, concentrations of certain ions). Many processes are actually biogeochemical and functions of the microflora must be taken into account. It may therefore be expected that progress will be made in our knowledge of the geochemistry of salts, including the physical chemistry of electrolyte solutions, soil physics and especially the transport of solutes and accessibility of sites to air and solutions that affect waterlogging (gleyic properties) as well as microbiology and biology of saline soils.

Further Improvements in Modes of Management or Restoration of Saline Soils that Take into Account 'Lessons from the Past'

Recent progress in modelling the phenomena of transport of water and solutions in soils can evidently be applied to the particular case of saline soils. Many attempts have been made in this direction, especially in the USA. However, in spite of their complexity, deterministic models do not take into account the entire diversity of agricultural practices and their interactions (soil tillage,...) nor the specific features of each mode of irrigation (submersion, spray, drip,...). As for stochastic models, they probably underestimate the deterministic character beyond the considerable apparent variability of salt accumulation in the phenomenon of salinisation (Rhoades, 1994).

It is therefore worthwhile to simultaneously improve management methods based on the leaching fraction (LF) concept defined above and not to neglect practices which indeed often present an empirical character and remind us of 'culinary recipes'. The following aspects are of particular interest:

— maintaining phreatic water in irrigated areas at an adequate depth (2.5 to 3.5 metres) so that the phenomena of capillary rise and waterlogging are reduced;

— placing, whenever possible, a mulch on the soil surface to reduce evaporation;

— maintaining good hydraulic conductivity so that the transport of salts to a depth beyond the root zone is possible;

— adapting the procedures to climatic conditions: for example, ensuring that soils are moist at the beginning of winter so that the precipitations which follow may have a good leaching effect on the salts; or even reserving the use of more saline waters for periods when the crops are at a stage of maturity, because they are more tolerant at that time;

— mixing waters of different salinities in the management of large areas;

— and certainly carefully analysing at the level of catchment areas the origins of salts (sources) applied in cultivated systems.

All these practices, provided they are adapted to the local context, result from human beings having taken 'lessons from the past' into account and do not conflict with recent progress in modelling.

CONCLUSION

In view of the threat to agricultural production as a result of increase in saline soils, considerable flexibility needs to be adopted. Some authors (Szabolcs, 1994) consider that this challenge would be even more serious if global warming and a rise in sea level were confirmed. New areas, especially around the Mediterranean, would then be affected.

From this point of view, genetic research on improving the tolerance of plants to salts (Meiri, 1994) should certainly be further developed. This approach is necessary but probably insufficient. New ways have to be found. The biological aspects (macro and micro) have been much neglected, especially in the functioning of saline soils. Their implications in the principal biogeochemical cycles (nitrogen, carbon, phosphorus, sulphur) are, however, decisive (Dellal and Halitim, 1992). The role of organic matter on the development and consequences of salinisation and especially sodisation is not well known, however.

But whatever the skill with which drainage waters are managed and subsequently recycled, a time will come when these waters can no longer be used. This leads to the idea of 'ultimate wastes'. What should be done with these waters? Evaporating them in ponds reserved for the purpose is one possible method, but not without endangering the environment of such ponds.

The social parameter will remain decisive. People have often been confused by major installations (dams on large rivers), which have certainly increased the available land and water resources, but farmers were not ready to effectively manage the resources made available to them in this manner.

More generally, although in this book emphasis has been given to salinisation associated with irrigation, it should be kept in mind that other management practices such as encroachment of primary forest (for example in northern Thailand), overgrazing or merely changes in the cropping system, may also lead to the same effects. Salinisation is therefore one of the most significant indicators of poor management in rural areas by human beings.

Glossary

Alkalinisation	Increase in pH level of a soil subsequent to accumulation in it of salts such as sodium carbonate.
Salinisation	Accumulation in soils of highly soluble salts, chlorides and sulphates of sodium or, in particular, of magnesium. This accumulation hampers development of plants that find it more difficult to draw the water they need because of a change in osmotic potential of the soil water.
Sodisation	Fixation of sodium on the adsorption complex of soils. Sodisation is often observed by a degradation in the physical properties of particular soils (for example, tendency of the soil to become impermeable).

References

Al Droubi A. 1976. Géochimie des sels et des solutions concentrées par évaporation. Modèle thermodynamique de simulation. Application aux sols salés du Tchad. Sciences Géologiques, Mémoire, 46. Strasbourg, 177 pp.

Al Droubi A, Cheverry C, Fritz B, Tardy Y. 1976. Géochimie des eaux et des sels dans les sols des polders du lac Tchad: application d'un modèle thermodynamique de simulation de l'évaporation. Chemical Geology 17: 165-177.

Al Droubi A, Fritz B, Gac JY, Tardy Y. 1980. Generalized residual alkalinity concept; application to prediction of the chemical evolution of natural waters by evaporation. Amer. J. Sci., 280: 560-572.

Bernstein L. 1975. Effects of salinity and sodicity on plant growth. Annu. Rev. Phytopathol. 13: 295-312.

Bertrand R, N'Diaye M, Keita B. 1994. L'alcalinisation/sodisation, un danger pour les périmètres irrigués sahéliens. Sécheresse 3: 161-171.

Borlaug NE, Dowswell CR. 1994. Feeding a human population that increasingly crowds a fragile plant. 15th World Congress of Soil Science, Acapulco (Mexico), July 1994. Keynote lecture, 15 pp.

Bourrié G. 1975. Relations entre le pH, l'alcalinité, le pouvoir tampon et les équilibres de CO_2 dans les eaux naturelles. Science du So 3: 145-159.

Bourrié G, Lelong F. 1994. Les solutions du sol. In: Traité de Pédologie. Bonneau M, Souchier B (eds.). Masson, Paris (2nd ed.).

Bresler E. McNeal BL, Carter DL. 1982. Saline and Sodic Soils. Advanced series in agricultural sciences. Springer-Verlag, Berlin, 236 pp.

Cheverry C. 1974. Contribution à l'étude pédologique des polders du lac Tchad. Dynamique des sels en milieu continental subaride dans des sédiments argileux et organiques. Thése Université Louis Pasteur, Strasbourg, 275 pp.

Cheverry C, Robert m. 1993. Salure des sols maghrébins: Influence sur les propriétés physiques et physico-chimiques des sols. Répercussion des modifications de ces derniéres sur la fertilité, notamment azotée. Rapport final du contrat C.E.E. STD TS2-108-F, 34 pp.

Daoud Y, Robert M. 1992. Influence of particle size and clay organization on hydraulic conductivity and moisture retention of clay from saline soils. Appl. Clay Sci 6: 293-299.

Dellal A, Haltim A. 1992. Activités microbiologiques en conditions salines: cas de quelques sols salés de la région de Relizane (Algérie). Agricultures 1: 335-340.

Gac JY. 1979. Géochimie du bassin du Lac Tchad. Bilan de l'altération, de l'érosion et de la sédimentation. Thèse, Université Louis Pasteur, Strasbourg, 249 pp.

Hensley M. 1969. Selected properties affecting the irrigable value of some Makatini soils. MSc Agric. Thesis, University of Natal, Pietermaritzbourg, S. Africa.

Job JO. 1992. Les sols salés de l'oasis d'El Guettar (sud tunisien). Thèse de Doctorat, Université de Montpellier II (France), 149 pp.

Katerji N. 1995. Réponse des cultures à la contrainte hydrique d'origine saline: approches empiriques et mécanistes. C.R. Acad. Agric. Fr. 81 (2): 73-86.

Laudelout H, Cheverry C, Calvet R. 1995. Modélisation mathématique des processus pédologiques. Actes Editions, Rabat, 262 pp.

Le Brusq JY, Loyer JY, Mougenot B, Carn M. 1987. Nouvelles paragenèses à sulfates d'aluminium, de fer, et de magnésium, et leur distribution dans les sols sulfatés acides du Sénégal. Science du Sol. 25: 173-184.

Maas EV, Hoffman GJ. 1977. Crop salt tolerance—current assessment. J. Irrig. Drain. Div., Proc. Am. Soc. Civil Eng. 103: 115-134.

Matsumoto S, Zhao Quiguo, Yang J, Zhu S, Li L. 1994. Salinization and its environmental hazard on sustainable agriculture in East Asia and its neighbouring regions. 15th World Congress of Soil Science, Acapulco (Mexico), July 1994, vol. 3, symp. A, pp. 236-255.

Meiri A. 1994. Tolerance of different crops to salinity conditions in soils. 15th World Congress of Soil Science, Acapulco (Mexico), July 1994, vol. 3, symp. A, pp. 320-331.

Meybeck M. 1984. Les fleuves et le cycle géochimique des éléments. Thése, Université Pierre et Marie Curie, Paris 6, 530 pp.

Meybeck M. 1987. Global chemical weathering of surficial rocks estimated from river dissolved loads. Amer. J. Sci. 287: 401-428.

Miyamoto S, Muller W. 1994. Irrigation with saline water: certain environmental implications. 15th World Congress of Soil Science, Acapulco (Mexico), July 1994, vol. 3, symp. A, pp. 256-277.

Montoroi JP. 1994. Dynamique de l'eau et géochimie des sels d'un bassin versant aménagé de Basse Casamance (Sénégal). Conséquences sur la gestion durable de l'écosystème de mangrove en période de 'sécheresse. Thèse, Université de Nancy 1, 349 pp.

Oster JD. 1994. Management of irrigation water and its ecological impact. 15th World Congress of Soil Science, Acapulco (Mexico), July 1994, vol. 3, symp. A, pp. 332-345.

Oster JD, Rhoades JD. 1975. Calculated drainage water composition and salt burdens resulting from irrigation with river waters in the western United States. J. Environ. Qual. 4: 73-79.

Pitzer KS, Simonson JM. 1986. Thermodynamics of multicomponent, miscible, ionic systems: theory and equations. J. Phys. Chem. 90: 3005-3009.

Qiguo Zhao. 1994. Foreword. 15th World Congress of Soil Science, Acapulco (Mexico), July 1994, vol. 3, symp. A, p. 234.

Rhoades JD. 1994. Soil salinity assessment: recent advances and findings. 15th World Congress of Soil Science, Acapulco (Mexico), July 1994, vol. 3, symp. A, pp. 293-311.

Rieu M, Touma J, Gheyi HR. 1991. Sodium calcium exchange on Brazilian soils: modeling the variations of selectivity coefficients. Soil Sci. Soc. Am. J. 55: 1294-1300.

Summer ME. 1993. Sodic soils: new perspectives. Aust. J. Soil Res. 31: 683-750.

Szabolcs I. 1979. Review of Research on Salt-affected Soils. Natural resources research XV, UNESCO, 137 pp.

Szabolcs I. 1994. Prospects of soil salinity for the 21st century. 15th World Congress of Soil Science, Acapulco (Mexico), July 1994, vol. 1, pp. 123-141.

Tardy Y. 1993. Pétrologie des latérites et des sols tropicaux. Masson, Paris, 459 pp.

Vallès V, Packepsky I, Bourrié G. 1995. Caractéristiques agronomiques des sols salés méditerranéens Représentativité par rapport aux sols salsodiques mondiaux. In: Facteurs limitant la fixation symbiotique de l'azote dans le Bassin méditerranéen. Drev J (ed.). INRA Editions, Paris, Les Colloques no. 77.

Van Breemen N. 1976. Genesis and solution chemistry of acid sulfate soils in Thailand. Thesis. Centre for Agricultural Publishing and Documentation, Wageningen, The Netherlands, 263 pp.

CHAPTER 9

Erosion of Cultivated Soils by Water in Temperate Conditions

Y. Le Bissonnais, C. Gascuel-Odoux

THE RISKS

Soil erosion is the phenomenon of removal of material from the soil surface. If this occurs by water, it is known as water erosion; if by wind, as aeolian erosion. Erosion by water covers different types, such as coastal erosion, river erosion or that which is due to civil engineering activities. This chapter deals only with pluvial erosion, by far the most important in a rural zone with temperate climate.

As will be seen, pluvial erosion results from different processes depending on raindrops, sheet flow or concentrated flow in channels, be they discontinuous or a part of a hydraulic network. Pluvial erosion is highly variable from one point to another, or from one hydrographic basin to another. Fourmier, cited by Hénin (1978), reported that in some cases 90% of the solid material removed by erosion is from only 10% of the surface in a hydrographic basin with considerable disparities between basins; from 17 g m^{-3} y^{-1} for the entire Seine basin, it may be 15 kg m^{-3} y^{-1} in the fragile zones of the Durance basin, or about 1000 times more, and may represent the removal of a centimetre-thick layer every year. These overall estimates at the level of major catchment areas do not take into account the redistribution of soil within the basins that is often of primordial significance.

This problem of pluvial erosion is recognised as important and has long been studied in tropical regions, Mediterranean regions and montane areas, characterised either by harsh climates or steep slopes, and where additional problems involving degradation of plant cover (desertification, land clearings, ...) often occur. Erosion, on the other hand, assumes an unexpected feature in temperate regions and large farms of north-western Europe where the slopes are gradual with gradients of a few per cent, where rainfall is relatively light and of low intensity. For example, a daily rainfall of more than 40 mm has been recorded for just 3 to 6 days in 10 years in these regions, while in the Mediterranean French area such a heavy rainfall is recorded for 3 to 6 days every year.

It is to these less harsh contexts in the regions of north-western Europe—from the Paris basin east to the Ukraine, including the Belgian, German and Polish plains—that this chapter on water erosion is confined. The processes of erosion have been a cause for increasing concern, partly because of intensification of agriculture. In this context, Bollinne (1982) estimated that soil loss may be as much as 3-3.5 kg/m^2 and that the entire agricultural loss in a three-year crop rotation due to erosion is 3 to 5% of the crop value. Traditional approaches to erosion do not appear pertinent in these contexts, only recently studied. The state of the soil surface—an interface par excellent—plays a major role in the formation of the processes of runoff and water erosion.

Regions affected by the problem of water erosion in north-western Europe generally meet three types of criteria:
- Geomorphological criteria: soft geological substrates such as limestone are extremely vulnerable to such erosion; they result in a slightly undulating relief with long gentle slopes and no obstacle to surface flows, mechanisation or increase in plot size.
- Hydrological and pedological criteria determine the initial hydric conditions of the soil profile before rain storms and the conditions of transport during the storm; silty layers, given their low structural stability, are susceptible to beating down and compaction by agricultural machinery. The hydraulic conductivity of the first few millimetres of soil and the first 20 centimetres is often greatly reduced by a crust and soil compacted by agricultural machinery.

- Criteria associated with land use: consolidation of holdings and reduction in the area of grasslands in favour of more intensive systems of production increase the likelihood of a surface runoff. The number of passes of agricultural machinery and the duration of periods when there is limited plant cover increase, while finer working of the soil usually reduces roughness of the surface.

- Increasing concern for soil erosion in temperate regions is due on the one hand to increase and frequency of damage and, on the other, to the need for taking into account many long-term damages, especially those related to the quality of waters and soils. Damages differ depending on localisation:

- On the slopes they are suffered mainly by farmers. These are notably, from higher to lower levels, uprooting of plants and seedlings, creation of gullies that obstruct agricultural intervention, and burial of plants and seeds by deposits. In the longer term, removal of the finer elements and organic matter continues along the slope and contributes to a differentiation of the physical and chemical properties of the surface horizons, depending on their disposition on the slope. Apart from the damage associated with erosion processes, while redistributing the water along the slopes the runoff carries away plant treatment products and fertilisers, combined in part with the transported soil particles. The spatial distribution of the water reserve in the soil or of the fertilisers is, for example, altered by this redistribution of the water on the surface. Phenomena of plant toxicity may also appear at the lower end of the plot.

- At the bottom, damage is suffered mostly by regional co-operatives. These damages may be both sudden and spectacular, such as sand or mud blanketing the roads, blockage of rain-water harvesting networks, and even road cave-ins. The quality of surface waters may be drastically degraded. Increase in turbidity of these waters and mobile phosphorus content fixed on the particles may contribute to chokage of watercourses, their eutrophication and reduction in their quality.

Forecasting the extent, occurrence and localisation of the damage caused by water erosion of soils is extremely difficult due to the many factors involved, their interaction, the several levels of space and time in which they intervene and lastly, the non-linearity of the effects with respect to variations in these factors. For example, from the planting of a crop to changes in the surface state as and when climatic events occur, i.e., changes in the plant cover, surface roughness and morphological and physical properties of the very first centimetres of the soil, are relatively continuous. But for identical climatic events the responses are extremely variable depending, for example, on the moisture content of the soil and thus the series of preceding rainfalls.

Research on water erosion in various regions of north-western Europe is presently directed towards analysis of the processes and factors involved, investigation of the units of space and time during which they are active (known as functional units) and studies on the units of space and time in which these processes can be acted upon (known as 'decisional' units).

PRINCIPAL PROCESSES

The driving force in water erosion is the surface runoff. Runoff and erosion are the results of complex interactions between several factors: rainfall, soil type, land use and relief features, which interact in a series of processes that can be described in five stages (Fig. 9.1).

Infiltration, Surface Detention and Runoff

Rain-water infiltrates the soil upon reaching the surface. It is stored in the soil pores and redistributed in the profile in accordance with its physical properties. Two mechanisms may obstruct this infiltration. In the first and more common type, the intensity of rainfall is greater than the infiltration capacity of the soil surface. The hydraulic conductivity in saturated condition at the soil surface constitutes a limiting factor for infiltration. In the second mechanism, the infiltrated water may saturate the soil profile when the volume of rain exceeds its holding capacity. In this case it is the properties of the soil water reserve that limit infiltration. This happens in particular when the soil is not very thick and rests on a substrate of very low permeability.

The water that does not infiltrate is initially stored in depressions on the soil surface created by surface roughness. When the non-infiltrated volume exceeds the detention storage of the surface, runoff takes place.

The capacity for infiltration and the capacity of storage in soil and at the soil surface depend on soil type, initial moisture conditions, soil surface structure and plant cover, which undergo changes in time and during the course of rains.

Rains can degrade the structure of the soil surface, closing soil surface porosity and decreasing roughness, which greatly reduce the soil's capacity for infiltration. Soils which are mainly silty have a low aggregate stability that leads to soil crusting due to the effect of rain; such crusts are serious obstacles to rain-water infiltration. Its infiltration capacity may be

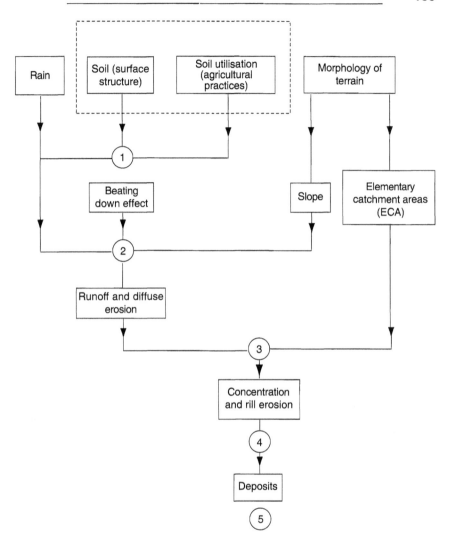

Fig. 9 1   Processes of runoff and erosion on silty layers in northern Europe (after King and Le Bissonnais, 1992).

reduced to values of a few mm h^{-1} when these crusts reach their final stage of development (Fig. 9.2). This degradation of the soil's surface is even more rapid when agricultural practices lead to small size of aggregate and clods and when soils are bare. Plots prepared in this manner constitute 'potential runoff' surfaces.

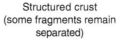

Structured crust
(some fragments remain
separated)

Sedimentary crust
(smoothing of the
surface)

Phase 0.	Phase 1.	Phase 2.
Initial fragmented porous and friable state after soil-working. Infiltration possible 30 to 60 mm h^{-1}	Closure of surface due to 'splash' effect. Infiltration possible 6 to 2 mm h^{-1}	Sedimentation in puddles. Infiltration possible 1 mm h^{-1}

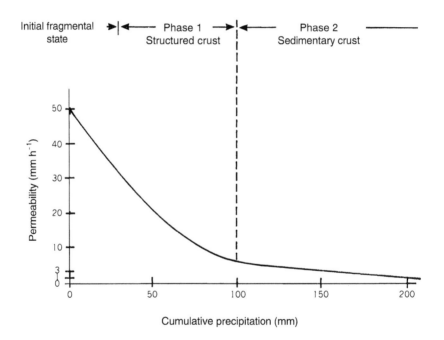

Fig 9.? Kinetics of reduction in soil permeability depending on the extent of beaten crust (after Boiffin, 1984).

Increase in soil density due to the effect of agricultural machinery also reduces this capacity for infiltration and storage. In other soils the obstacle to filtration may be a plough pan or an underlying horizon that is more clayey and impermeable. In such cases, if the slope allows, a subsurface runoff may occur at the level of the impermeable horizon before the holding capacity of the surface horizons is exceeded. This subsurface runoff is then likely to appear at the surface if there is a change in slope or a reduction in thickness of the surface horizons.

Processes of Detachment and Spurting: Effects of Raindrops

Raindrops have three effects on the soil surface: on the one hand, they break up clods and aggregates by various mechanisms, namely bursting, physicochemical dispersion, microcracking by differential swelling and mechanical disaggregation; on the other hand, displacement occurs through spurting of the separated fragments.

These processes together constitute what is referred to as 'splash' and result in a structural degradation of the surface called 'beating down'. There are several stages in this degradation of the surface structure by rain (Fig. 9.2): from an initial fragmentary state as a result of tillage, clods are broken up; elementary particles and microaggregates are detached and displaced by raindrops, and then collect and join together to form, in the first phase, the so-called structural crust. When there is an excess of water on the soil surface, the soil particles become suspended and separate. Subsequently, two things may happen to them: they may be either integrated into the crust structure as deposits producing the so-called deposit crust (especially when the particles are found in microdepressions on the surface) or are displaced on the soil surface due to the combined effect of spurting by raindrops and the surface runoff when it begins. It is this displacement of particles that is responsible for non-point erosion.

Diffuse Erosion: Effect of Sheet Flows

Sheet flows on relatively gentle slopes cause a diffuse slope erosion that progressively erodes the soil surface. This erosion is often neglected because per se not very spectacular and difficult to evaluate quantitatively. But recent studies have shown that volumes of several tens of $m^3 ha^{-1}$ may be displaced by this process of diffuse erosion in a plot during a single rain storm. The soil surface is then often marked by erosion rills which are small discontinuous furrows a few mm deep. This insidious type of erosion that displaces large quantities of earth may, in fact, cause rapid collapse of management installations such as water storage basins and increase their maintenance costs. Further, the water and combined salts may cause pollution if this happens just after a plant protection treatment or fertiliser application.

If the slope steepens, deeper furrows of a few centimetres may develop at fairly regular intervals. These furrows often follow tyre tracks or tillage lines. This amounts to furrow-interfurrow erosion systems combining diffuse and linear erosion.

Linear Erosion: Role of Morphology of Catchment Area

Depending on the form of catchment areas, there may be a concentration of runoff with local formation of an often considerable water flow that causes linear erosion. The tractive force of the runoff is then sufficient at the points of concentration in the runoff or along the principal axis of drainage in the elementary catchment area to cause the formation of furrows (incisions centimetres or decimetres) and gullies (cuts decimetres to metres deep). The latter, often spectacular, may prevent movement of agricultural machinery and lead to mudslides at the bottom (toeslope).

The factors involved in deepening and propagation of these gullies are still not well known, but it is certain that the mechanical cohesion of various soil horizons plays an important role. Actually, if during its formation a gully reaches a more friable horizon, there is further deepening and regressive erosion by a collapse and reduction of the cuts upwards from the lower side. The local slope and size of the potentially erodable collection network are also important factors.

Deposits and Mudslides

The particles displaced by concentrated or diffuse erosion may be deposited at an obstacle or where there is a change in the state of the surface or the vegetation, or even a reduction in gradient, which reduces the effect of the runoff. This results in the formation of a mudslide, generally at the lower end of the plot or at the toeslope. These deposits may create problems when they occur on agricultural land (burial of crops, ...) but the damages are less than when the displaced particles reach the road and the hydraulic network, causing mostly damages for communities and for society in general.

The spatial separation of the different processes described above leads to characterising the erosion factors and assessing the risks at the level of the elementary catchment area, constituting the major spatial unit for integrating all the processes involved.

SPATIAL AND TEMPORAL FUNCTIONAL UNITS

An elementary catchment area is an area of land from several hectares to several tens of hectares that collects the surface waters going to the same watercourse. It is indicated by a point on the watercourse, the outlet, and by a line dividing the surface waters which, at each side of the outlet, follows the line of the steepest slope to the crest, thereby separating a particular basin from contiguous ones (Fig. 9.3a). This topographical definition is highly pertinent for surface transports. It is presently easy to use in an automatic manner with numerical terrain models (NTM), altimetric bases, generally defined on regular square meshes made from altimetric charts or by photogrammetry (on spot images or aerial photographs).

The catchment area is an open system; it has different external exchanges; it receives flows, mainly on the soil surface, from precipitations or wind; and produces flows, either in vapour form through the soil surface and plants or in liquid or solid form at the outlet. Within there is storage, redistribution and transformations, either on the soil surface or in the soil and subsoil. The catchment area is thus the spatial unit where balances are established—water balance, erosion balance, geochemical balance—and in which the processes of surface transport are studied.

In the case of water erosion, the catchment area is divided according to the various processes of particle transfer (Fig. 9.3b):

- the slope, representing almost the impluvium, is the zone of erosion and runoff production;

- the bottom domain is, first of all, the zone of concentrated erosion

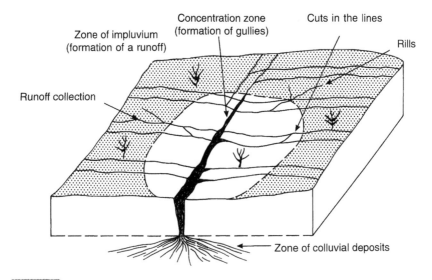

fig 9 3a Zones differentiated according to runoff and erosion in an agricultural catchment area (after Auzet, 1987).

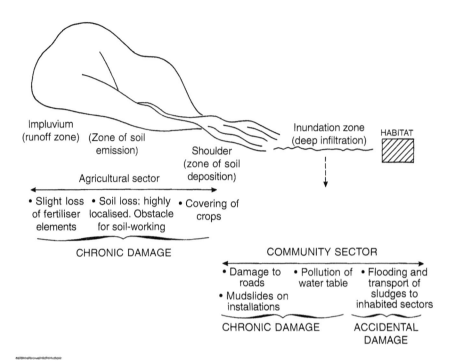

Impluvium
(runoff zone)   (Zone of soil
                  emission)

Shoulder
(zone of soil
deposition)

Inundation zone
(deep infiltration)   HABITAT

Agricultural sector

• Slight loss   • Soil loss; highly   • Covering of
  of fertiliser   localised. Obstacle   crops
  elements   for soil-working

CHRONIC DAMAGE

COMMUNITY SECTOR

• Damage to   • Pollution of   • Flooding and
  roads   water table   transport of
• Mudslides on   sludges to
  installations   inhabited sectors

CHRONIC DAMAGE   ACCIDENTAL
                  DAMAGE

Fig. 9.10   Various zones in a catchment area and types of damage (after Papy and
Douyer, 1991).

where considerable damage due to formation of gullies occurs;
several farmers whose plots lie in this zone suffer such damage;

• next is the zone of deposition, a flat domain, an area where crops
  are buried, causing losses to farmers and more particularly to
  communities through damage to roads and houses.

The importance of higher zones from where runoffs are produced is
thus evident with reference to damage from erosion in the lower regions.
One strategy in the study of erosion risks is to compare the modality of
spatial distribution of the factors causing erosion, estimated at the level of
catchment areas, with the effects observed in the lower areas of these
basins. The use of Geographic Information Systems (GIS) helps in this
approach. For example, it is possible to estimate the spatial distribution of
soils susceptible to being 'beaten down', soils of steep slopes, and those
with a limited plant cover,.... Similarly, the percentage of loss can be
estimated based on the extent of the network of gullies. Such analyses,
however, show that none of these factors is, by itself, decisive; on the

contrary, it is the geographic extent of spread of these zones that is determinative.

In a catchment area, other spatial units are also of importance not merely for purposes of evaluation, but for analysis of the origin of erosion, estimation of runoff and diffuse erosion, prediction of water flow on the soil surface and sites of formation of furrows. In plots, it is the role of the surface roughness created by soil-working and the topography: on the slope it is the study of the role of the spatial organisation of plots on it and the aspect of the slope in relation to the flow network. The following questions must be answered: is there a topographical threshold at which the water does not follow the microrelief at the soil surface but rather the line of the steepest gradient? Is the runoff on a particular plot likely to infiltrate during its passage from a depression on the slope the surface of a series of plots?

The agricultural year is also a functional and integrated unit of time for evaluations because it takes into account the overall dynamics of the soil surface characteristics within a catchment area. For example, a study of several catchment areas in the region of Caux (Fig. 9.4) revealed a cyclic change in the percentage of soil surface likely to be eroded during the course of an agricultural year. Every year more than 90% of the area in

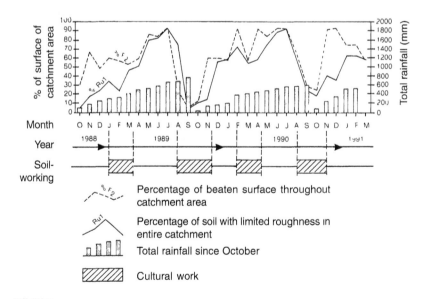

Fig 9.4 Monthly change in the surface characteristics of soil in an elementary catchment area in the region of Caux based on cumulative rainfall and tillage (after King and Le Bissonnais, 1992).

the catchment has a soil surface that is porous and very rough when agricultural work is carried out during the month of September. Later, due to rains, this surface becomes degraded and covers more than 70% of the area in the catchment for several months. It has also been noted that there is an overall change in catchment area during an agricultural year despite considerable variation in surface characteristics of the plots at a given time.

Likewise for space. There are other important temporal units for understanding the origin of the processes of water erosion. These are notably the series of earlier rains and the characteristics of each storm which constitute the functional units of time that interact and determine the extent of water erosion as explained below.

FACTORS OF WATER EROSION

Rain

Rain is obviously the principal agent in water erosion. What is known as the erosivity of rain is its ability to trigger erosion. Erosivity depends mainly on the intensity of rainfall or the kinetic energy resulting directly from it. But because of the many interactions that occur between the rain factor and other factors, it is somewhat difficult in practice to forecast the effect of a particular storm event on erosion.

Actually, according to classic criteria for evaluating erosivity of rain, the oceanic climate over a large part of France could be characterised by a very low erosivity because of the moderate amount and intensities as well as the uniform distribution of rains during the year, which contradicts erosion features that are actually observed.

To properly understand the role played by the rains and the effect of their characteristics, it is necessary to distinguish clearly between the characteristics of each storm and the overal rainfall characteristics in a season and to incorporate them in the overall scheme of the erosion processes just described.

The cumulative rainfall from the last tillage to a rain event will determine the level of structural degradation on the soil surface: the rainfall of a few days earlier will determine the moisture content in the soil on the date of the event, while the characteristics of the shower itself will determine the hydrodynamic behaviour of the soil during this shower interacting with all the other parameters, including those that depend on the earlier rains. Statistical studies have revealed that considerable erosion occurs in situations combining heavy rains, prolonged heavy rainfall during the five previous days, and significant cumulative rainfall since the last tillage. No single one of these parameters suffices to cause erosion, unlike a tropical or Mediterranean type situation in which erosion may be associated with the characteristics of a single shower.

Soil Properties

Soils affect the runoff process mainly as a result of their permeability and aggregate stability. The permeability of a soil depends on the series of horizons and their physical properties. Generally, the more the successive horizons with contrasting physical properties, low hydraulic conductivity and low water-retention capacity, the more limited the infiltration. This applies to clay soils for example.

The most determinative properties, however, are those of the surface horizon which is directly subject to the effect of atmospheric agents. Whatever the properties of the underlying horizons, they often act as the main obstacle to infiltration when their structure is degraded by rain and a crust is formed due to the impact of rains.

The aggregate stability of this horizon is therefore a determinative parameter. It depends essentially on the texture, organic matter content and physicochemical properties of soils (arrangement of cations, pH, ...). Soils with a large amount of organic matter generally have a good aggregate stability but when used in agriculture there may be a progressive reduction in the organic matter content. In soils of the Paris basin, for example, this content is often between 1 and 2%. In such conditions silty soils, found over a large part of the zones with large-scale farming in north-western Europe (leached brown soils), are highly susceptible to the 'beating down' phenomenon, especially when their clay content is low. Calcic or calcareous soils are generally less susceptible to the battering effect as are soils with a coarse texture—less well structured but through which infiltration is possible.

Agricultural Practices

Under this term various kinds of human intervention in the environment can be included, at different time scales.

In the long term it is the manner of soil use, with spatial distribution between annual crops, grasslands and wooded zones or in fallow areas, with the possible intervention of consolidation of holdings that may greatly affect the landscape and influence the phenomena of runoff and erosion by increasing the size of agricultural plots and eliminating the obstacles to runoff such as hedges between plots, for example.

In the medium term, it is the agricultural system which implies an assessment of crop rotations over several years. Changes in the cultural systems because of technical or economic factors may induce overall changes in runoff by, for example, increasing the ratio of areas with winter sowing compared to those of grasslands.

Lastly, in the short term, by choice of agricultural practices, types of tools or dates of intervention that help modify the surface characteristics of the soil at a given time, the farmer is mainly responsible for the risks of runoff and erosion. The effectiveness of these choices is quite often restricted, however, by constraints which may be of a technical nature (need for breaking up the soil to ensure good germination), economic, climatic, or due to the agricultural programme itself. An effective compromise has to be reached.

Morphology of the Terrain

The morphology of the terrain affects erosion in two ways: on the one hand, by the slope which determines the flow regime of the waters at a given point on it and, on the other, by the limits of the elementary catchment areas and other resultant characteristics which indicate, at a given point on the catchment, the extent of the areas that contribute to surface flows.

Although the slope is *a priori* a favourable factor for runoff and erosion, its effect is not easy to indicate on soils with large farms that are generally located on gentle slopes. Intensity of runoff appears to be associated more with characteristics of the soil surface than with the slope. On the other hand, for an equivalent runoff, formation of furrows may be associated with the slope at a threshold of about 5%. Further, the intensity of the slope along with direction and type of tillage, determines the mode of flow on the surface of cultivated plots. It is generally considered that on steep slopes water flows along the line of the steepest gradient. This is not necessarily the case when the slope is gentler: then water may flow in the direction of tillage. The overall result of the direction of water flow finally depends on the complex interactions between the slope, direction of tillage in relation to line of the steepest slope, surface roughness associated with tillage and intensity of runoff.

The general morphology of the catchment area also has an effect on other morphometric characteristics of catchments that influence water erosion. These are the overall size of the catchment in particular, its form—concavity and convexity—and length of the slopes, which eventually help to estimate the areas that feed runoff at each point in the catchment area. Here again, these actually result from complex interactions between the slope and successive characteristics of soil surface, and soils that depend respectively on the layout of the plots and the type of soil in the slope.

In contrast to erosion processes involving a series of factors at different space-time levels, as already mentioned, strategies of anti-erosion measures are eventually reflected in the decisional units.

DECISIONAL SPATIAL AND TEMPORAL UNITS AND STRATEGIES FOR ANTI-EROSION MEASURES

It is the farmer who makes decisions regarding use and management of the plots which constitute his farm. Several anti-erosion measures may be considered at this level with a view to reducing generation of runoff by changes in agricultural practices in the plots, or to channellise erosion by a network of drains, or to reduce its abrasive effect on the soil by packing or planting grass on thalwegs. Though these techniques have generally proven effective, it is often difficult to apply them because of sociological or economic constraints.

Take, for example, packing the soil at the bottom of a thalweg to increase resistance to furrowing. A comparison of catchment areas subjected to the technique with other such areas revealed its positive effect. Nevertheless, its adoption depends on the spatial organisation of the plots in a particular farm covering several catchment areas. A farmer who has plots at the higher level from where runoff begins, and plots at the lowest level that suffer damage, will be aware of the problem of erosion. He will have a better idea of the implementation and impact of this technique than a farmer whose plots are dispersed and isolated in several catchment areas. The structure of the habitat, depending on whether it is grouped or dispersed, is a factor for regrouping plots around farms. Analysis of farming structures is therefore essential before considering any technical proposition with respect to anti-erosion measures.

Lastly, many decisions are made by local communities such as the commune or department, or international bodies such as the Commission for European Communities. In such cases available data are often imprecise or unadapted to functional units. For example, agricultural statistics are prepared by a canton, soil analyses are classified by a commune, while evaluation of risks of erosion is carried out mainly at the level of a catchment area. Analysis of the usefulness of the chosen indicators at the decisional level is therefore equally essential. Elaboration of geographic reference databanks, combined with utilisation of the MNT, should help in systematic verification of the value of these indicators in future.

CONCLUSION

The problem of water erosion in the regions of north-western Europe has been dealt with in this chapter: the soil and climatic conditions and considerable human activity in the environment in these regions have led to notable weightage of the characteristics of the soil surface in the origin of erosion processes. These characteristics are extremely variable in time and space. Decisions have to be made as to which processes intervene in modifying the characteristics of the soil surface, in which periods they intervene and to which areas they correspond. In other contexts, assessments may differ. Thus, in tropical situations, the characteristics of rainfall are of primary importance while the surface characteristics of the soil are less variable in time and space.

A study of water erosion helps us deal directly with the problem of transport of pollutants that move, for example, along with particles of phosphates and pesticides. In a more indirect way, assessments made for units of space and time within the problem of water erosion can be used for other problems such as transport of water and pollutants in general. The processes are obviously governed by mechanisms of an entirely different type: processes of convection, diffusion, adsorption depending on the level of transport of water and solutes. But the catchment area is also important as a primary functional unit, as are differentiation in space of a zone of emission on the slope and a zone of purification at the lowest levels, and differentiation in time of different phases during the course of the hydrologic cycle. It is also to be noted that there are similar decisional levels that determine spatio-temporal distribution of inputs and the conditions for their propagation on the slope.

Glossary

Aggregate stability	Susceptibility of soils to processes of disintegration. Various laboratory tests have been designed to evaluate it, which provide a description and classification of the structural behaviour of soils due to the effect of water.
Beating down	This results from disaggregation and consists of formation of particular structures on the soil surface, known as beaten crusts: we distinguish structured crusts which result from reorganisation of the surface structure and closure of surface porosity, and sedimentary crusts that result from successive deposits of sediments in puddles which appear after an excess of water.

Detention storage	The microtopography creates unevennesses and hollows that indicate the sites where puddles will appear before a runoff begins. The volume represented by all these puddles filled to maximum is termed detention storage.
Disaggregation	Process of fragmentation of soil that occurs due to various mechanisms and affects different levels of soil structure, from interactions between clay particles to soil clods. The main mechanisms in this break-up are physical and chemical dispersion, bursting due to trapped air, cracking due to swelling/shrinkage, and mechanical due to the impact of water drops.
Infiltration capacity	Also known as infiltrability, expresses the quantity of water which infiltrates per unit of time. It depends on the constituents and arrangement of soil porosity. It varies in time depending on the moisture conditions of the soil.
Non-point erosion, concentrated erosion	Erosion is a process of separation and transport of solid material. The term also refers to an assessment of matter removed from a unit of surface area. Transports, if associated with water, occur in the form of a sheet flow distributed almost uniformly over the soil surface, referred to as non-point erosion; flow in a localised manner, in drains or channels is known as concentrated erosion.
Runoff	Strictly speaking, runoff is a flow of water over the soil surface due to gravity. It is also termed surface runoff to avoid any confusion. In its broader meaning, hydrologists use the term runoff to describe the rapid flow seen during a flood. The word, therefore, refers to flow over the soil surface and through the soil as well.

FURTHER READING

Auzet AV. 1987. L'érosion de sols par l'eau dans les régions de grandes cultures: aspect agronomique. Ministères de l'environnement et de l'agriculture, 60 pp.

Auzet AV, Boiffin J, Papy F, Ludwig B, Maucorps J. 1992. Rill erosion as a function of the characteristics of cultivated catchments in the North of France. Catena, 20: 41-62.

Boiffin J. 1984. La dégradation structurale des couches superficielles du sol sous l'action des pluies. Thèse Doct. Ing. INAPG, 320 pp. + annexes.

Bollinne A. 1982. Etude et prévision de l'érosion des sols limoneux cultivés en moyenne Belgique. Thèse Univ. Liège, 356 pp.

Bresson IM, Boiffin J. 1990. Morphological characterization of soil crust development stages on an experimental field. Geoderma, 47: 301-325.

Gascuel-Odoux C, Bruneau P, Curmi P. 1991. Runoff generation: assessment of relevant factors by means of soil microtopography and micromorphology analysis. Catena, 4: 209-219.

Hénin S. 1978. *L'érosion liée à l'activité agricole en France.* Colloque sur l'érosion agricole des sols en milieu tempéré non méditerranéeen. Ed. Vogt, Université Louls Pasteur, Strasbourg.

King D, Le Bissonnais Y. 1992. Rôle des sols et des pratiques culturales dans l'infiltration et l'écoulement des eaux. Exemple du ruissellement et de l'érosion sur les planteaux limoneux nord de l'Europe. C.R. Acad. Agric., 78(6): 91-105.

Le Bissonnais Y, Renaux B, Delouche H. 1994. Interactions between soil properties and moisture content in crust formation, runoff and inter-rill erosion from tilled loess soils. Catena.

Monnier G, Boiffin J, Papy F. 1986. Réflexation sur l'érosion hydrique en conditions climatiques et topographiques modérées : Cas des systèmes de grande culture de l'Europe de l'Ouest. Cahiers ORSTON, Série Pédologie XXII (2): 123-131.

Papy F, Douyer. 1991. Influence des états de surface du territoire agricole sur le déclenchement des inondations catastrophiques. Agronomie 11 (3): 201-215.

Valentin C, Bresson LM. 1992. Morphology, genesis and classification of surface crusts in loamy and sandy soils. Geoderma, 55: 225-245.

CHAPTER 10

Sources of Soil Pollution

S. Martin

HOW DO TOXIC PRODUCTS REACH SOILS?

Through the Atmosphere

Apart from its gaseous components, the atmosphere contains particles, aerosols, which originate from various sources, both natural or associated with human activity.

Active volcanoes eject fine particles of rocks into the atmosphere. In arid and semi-arid regions, the wind erodes soils and every year in this manner the quantity of particles reaching the atmosphere is estimated at a billion tons; about 20% of these particles are transported over long distances. This is how sediments of the North Atlantic Ocean have, to a large extent, been formed from dusts from the Sahara. Moreover, oceans release a very large amount of particles into the atmosphere: when bubbles burst on the surface of the sea, fine droplets are carried by the wind and after the water in them has evaporated, the salts crystallise and form aerosols (Schlesinger, 1991).

At a global level, in the case of some metals, the emissions associated with human activity are equal to or even exceed those from natural sources (Table 10.1).

Table 10.1   Various emissions in the atmosphere in 1983 of trace elements due to human activity (Nriagu and Pacyna, 1988) and annual global emissions into the atmosphere of trace elements of natural origin (Nriagu, 1989). Median values provided by various estimates are given.

10^9 g y^{-1}	As	Cd	Cr	Cu	Hg	Ni	Pb	Se	Zn
Total human activity	18.82	7.57	30.48	35.57	3.56	55.65	332.3*	3.79	131.9
Total from natural sources including	12	1.3	44	28	2.5	30	12	9.3	45
erosion	2.6	0.21	27	8	0.05	11	3.9	0.18	19
volcanoes	3.8	0.82	15	9.4	1	14	3.3	0.95	9.6
natural forest fires	0.19	0.11	0.09	3.8	0.02	2.3	1.9	0.26	7.6
ocean fogs	1.7	0.06	0.07	3.6	0.02	1.3	1.4	0.55	0.44
biogenics (pollens, waxes, spores, leaf debris)	3.86	0.24	1.11	3.31	1.4	0.73	1.74	7.41	8.1

*including 248 × 10^9 g y^{-1} for transports

The majority of these particles with a diameter of more than 20 μm rapidly settle down after their emission. It is estimated that only particles with a diameter of less than about 10 μm may remain in the atmosphere for more than a few days and be carried to distances of more than 1000 km. Particles with a diameter of less than 1 μm remain in the air due to Brownian movement, acceleration of which in this case is more than that of gravity. Sooner or later, however, they all return to the soil by precipitation or after being agglomerated (Puxbaum, 1991; Knap, 1990).

Long-distance Transport

A study of lead in the snow and ice of the Antarctic and Greenland revealed that the entire troposphere has been contaminated by lead emitted as a result of human activity (Table 10.2).

Table 10 2   *Variations over the ages in lead concentration in snow and ice of the Antarctic and Greenland (after Boutron, 1988)*

$10^{-12} g$ of Pb g^{-1}	5000 BC	1980
Antarctic	0.4	2
Greenland	1	30

In the south-eastern region of France, at sites far from any incidental source of pollution and where the west winds that originate from the Atlantic Ocean predominate, it was possible to compute the increased content of metals in soils attributed to long-distance transport of aerosols (Table 10.3).

Table 10 3   *Atmospheric deposits on soils in Landes of Gascony (kg· ha⁻¹) (after Saure and Juste, 1994)*

	Cd	Cr	Cu	Ni	Pb	Zn
Atmospheric deposits	0.6	5	12	7	25	6

Rain-water and snow, even when there is no pollution, are always slightly acidic (pH: 5 to 6) because of dissolution of carbon dioxide and natural emissions such as sulphur from marine sprays and volcanoes in the atmospheric water. Human activities may considerably increase this acidity (Ramade, 1989). Thus the industrial and population concentration in the Rhine valley and Central Europe is such that, in the Vosges, rains and snows from the east and north-east are far more acidic (pH: 3.1 to 3.4) than those from the south-west (pH: 4.5 to 5.2). The acidity of these rains is not sufficient to cause direct damage to the leaves on trees, however. This direct damage appears only when the pH of the atmospheric water is less than about 2.6. Contrarily, acidic depositions on soils or plants, together with the natural poorness of some soils in the Vosges, result in mineral shortages that are very harmful to trees, especially a shortage of calcium and magnesium (Bonneau, 1992).

Measurements taken on acid brown soils in the Fougères forest, near the city of Rennes, have revealed an overall gamma radioactivity (electromagnetic vibration with a very short wavelength) of 638 Bq kg^{-1} (1 Becquerel = 1 decay per second). Most of this radioactivity is natural. The artificial radioactivity brought about by the atmosphere, represented by caesium 137, contributes only 2 Bq kg^{-1} to the total amount, a low value for France (Bourrié, 1993) (Table 10.4).

Table 10.4   *Gamma radioactivity of soils in the Fougères forest, near the city of Rennes, measured on 27 May 1986, one month after the Chernobyl accident: the only artificial radioelement is caesium 137 (Bourrié, 1993)*

	^{238}U Uranium	^{226}Ra Radium	^{210}Pb Lead	^{232}Th Thorium	^{40}K Potassium	^{137}Cs Caesium	Total
Bq kg^{-1}	39	37	37	39	484	2	638

Radionuclides of natural origin found in soils have been received from atmospheric fallouts associated with nuclear explosions since the beginning of the 1960s and the accident at Chernobyl on 20 April 1986, either through the atmosphere or with irrigation waters and routine wastes from nuclear installations. Presently in France, caesium 137 (half-life 30 years; contents between 1 and 190 Bq kg^{-1}), strontium 90 (half-life 28 years; contents less than 10 Bq kg^{-1}) and plutonium 238, 239 and 240 (half-lives 87 years, 2.4 × 10^4 years and 6.6 × 10^3 years; contents of plutonium 238 less than 0.1 Bq kg^{-1}; contents of the total 239 + 240 less than 0.5 Bq kg^{-1}) have been recorded. Other artificial radionuclides are present occasionally, such as caesium 134 (half-life 2.1 years; contents less than 5 Bq kg^{-1}). Radioactivity of soils has been the subject of a permanent study at the Institute for Nuclear Protection and Safety (INPS) and all results have been published (source: INPS, pers. comm.).

Short-distance Transport

Locally, emissions caused by man may be by far the main source of aerosols. Thus, in the Pas-de-Calais department, a metallurgical combine dealing in lead and zinc ores since 1920, has contaminated the soils in an area of about 70 km^2 (Godin, 1985; see Table 10.5).

Table 10.5 *Metal content of soils (mg kg^{-1}) in an agricultural plot located in the Pas-de-Calais department at 1200 m and 1500 m respectively from two metallurgical combines compared with (1) content in soils of an agricultural plot in the northern department, far from any occasional source of pollution, (2) soils in the Fougères forest in Brittany and (3) critical values that may not be exceeded, as defined in French law (Leprêtre, 1995; Leprêtre and Dessaint, 1988; Wopereis et al., 1988)*

	Metallurgical site (Pas-de-Calais)	Large farms (north)	Forest (Ille-et-Vilaine)	Critical values
Cadmium	8.9	0.2	0.05	2
Lead	531.5	56.4	21.4	100
Zinc	643	85.7	36.3	300

Vehicular traffic, because of the use of fuels and the wear of road, tyres and braking systems, is responsible for emission in the atmosphere of many contaminants: lead, cadmium, zinc,.... It is the main source for release of lead in the atmosphere. Contamination of soils along highways is now well known (Dumont and Vernette, 1974; Gaber, 1994; Harrison et al., 1981; Laxen and Harrison, 1977; Lebreton and Thévenot, 1992; Yassoglou et al., 1987) (Table 10.6).

Table 10.6 *Lead content in soils (0-15 cm) measured in 1992 along highway A36 that links Mulhouse and Belfort and which was opened in 1976 (after Gaber, 1994)*

	North			South		
Distance from highway (m)	10	5	2	2	7	15
Lead content (mg kg^{-1})	60	70	200	200	105	60

By Spread of Wastes

Urban sewage treatment stations treat waste waters from homes and industries connected with the sanitation network. This produces a treated water, acceptable to the natural environment into which it is released, and also a residue consisting of substances extracted from the waters during treatment known as 'treatment or sewage sludge'. This sludge is eliminated by incineration and disposal of its ashes either by dumping or spreading them on agricultural soils. The desire to reduce the use of fertilisers by recycling wastes, the increasing difficulty in arranging their disposal, the energy costs for incineration and treatment of fumes, led public authorities to encourage their use in agriculture. Thus, in 1991, of the 865,530 tons (dry matter) of sludge produced in France, 505,133 tons were used in agriculture. In 2010, the annual production of treatment sludge may be as much as 1,300,000 tons. Apart from the organic matter and fertiliser elements, such sludge often contains toxic substances, in particular metals in traces presently impossible to extract under acceptable technical and economic conditions. In these cases the agricultural channel has advantages that lead to its preference over competitors.

'The spread of sewage sludge on agricultural land cannot, however, be carried out without observing certain precautions that take into account the requirements for nitrogen and phosphorus by crops, and limiting the danger to health and the risks of supplying undesirable elements which may be contained in the sludges in excessive concentrations. French law has stipulated the conditions for utilisation of municipal treatment sludges, in particular a limitation on the concentrations of heavy metals in the sludges and in the soils that receive them, as well as a limitation on the quantities of sludges spread over the soil, depending on the heavy metal content in these sludges' (Table 10.7).

It may be noted that the mobility of a trace-metal element in the soil and its availability for plants depends on its *in situ* state (physico-chemical state, location in the soil), environmental conditions (pH, temperature, nature of the soil, presence of antagonistic elements, ...) and the nature of the plant: the quantities of trace-metal elements that enter plants and migrate to groundwaters cannot be predicted merely by a knowledge of their total contents in the contaminated soil and the critical thresholds represent only orders of magnitude vis-a-vis the actual risks of pollution.

The problem of elimination with respect to composts of domestic wastes arises in the same manner as for treatment sludge. It is estimated that of the 699,000 tons (dry matter) of composts of household wastes produced in France in 1991, 65% or 454,000 tons were used in agriculture (source: ADEME at Angers, pers. comm.) (Table 10.8).

Some watercourses must be cleared regularly. This involves removal of the sludge deposited at the bottom and sides of their beds. Until

Table 10 7 *Maximum limits (in mg kg^{-1} dry matter) of metals in sewage sludges and soils on which they are spread, prescribed by the French law (decree of 8 December 1997 and departmental order of 8 January 1998). Average metal contents in sewage sludges based on a survey conducted in 1992-94 on 237 urban treatment stations in France (Wiart and Verdier, 1995).*

	Maximum limits in sludges	Maximum limits in soils	Average content measured in treatment sludges
Cadmium	20	2	5.3
Chromium	1000	150	80
Copper	1000	100	334
Mercury	10	1	2.7
Nickel	200	50	39
Lead	800	100	133
Selenium	200	10	7.4
Zinc	3000	300	921
Chromium + copper + nickel + zinc	4000		

Table 10 8 *Average metal contents (in mg kg^{-1} dry matter) in composts of household wastes based on measurements obtained in France between 1981 and 1991 (source: I. Feix and Group on Municipal Wastes, ADEME at Angers, pers. comm.)*

Cadmium	Chromium	Copper	Mercury	Nickel	Lead	Zinc
3	128	250	3	71	434	915

recently, it was customary for such sludges to be spread on agricultural lands near these watercourses because they were considered to have the qualities of fertilisers. It was progressively noted, however, that there were trace-metal elements in the sludges due to poorly controlled urban and industrial releases. The precautions to be taken for spreading these sludges must be more rigorous than for sewage sludges since the quantities of sludge extracted from watercourses per hectare of a plot located near a stream may be as much as several hundred times the amounts prescribed for treatment sludges. The problem is particularly acute in northern France, where agricultural, industrial sites and urban zones are crowded together. Thus the Hôpitol stream in the city of Beuvry has, in the past, received the effluents from a galvanisation factory. The sediments are still contaminated (cadmium: 30 mg kg^{-1}; zinc: 6550 mg kg^{-1}) and the fields near the stream bear the stigma of the sludge spread on them (cadmium content in the soil: 16.5 mg kg^{-1}; zinc: 1100 mg kg^{-1}). In the years to come, the quantity of sediments to be extracted from watercourses in a province of northern France has been estimated at 230,000 t dry matter per annum (compared to 35,000 t sewage sludges produced in this province). Only about 10% of these sediments can be spread under permissible conditions. Storage of the remaining 90% in dumps is faced with operating costs and the difficulty of finding such sites. The problem has yet to be satisfactorily resolved and, for now, it is necessary to restrict clearing of polluted watercourses to the minimum possible extent (Six, 1992; Derville et al., 1994).

Copper as a growth factor and zinc, to avoid a skin disease, are incorporated in the feed for swine. Since these metals are not retained well in the animal, the slurry always contains a high content of them. They are found in the soil after spreading this slurry. In Brittany, a region known for intensive piggery, an average increase has been recorded per annum of copper content in soils of 0.37 mg kg^{-1} and of zinc 0.91 mg kg^{-1} (Coppenet et al., 1993).

With Supplies of Phosphates

Because of the presence of cadmium in most phosphate ores and the transfer of a large quantity of this metal to phosphate fertilisers made from them, particular attention has been paid for the last two decades to the consequences of using this type of fertiliser with respect to the cadmium level in cultivated soils. The phosphates used in European countries contain on average 6.6 g cadmium per 100 kg P_2O_5 (Hutton, 1983). A study conducted on standard reference plots existing since 1929 on the INRA estate in Versailles, has confirmed that the contribution of phosphate fertilisers represents an important part of the total cadmium pollution in soils which do not receive other spreads (sludges, composts, household wastes,...). The rest is due to atmospheric deposits, which have notably increased in the last thirty years (Tauzin and Juste, 1986) (Table 10.9).

Subsequent to Use of Pesticides (Mineral and Organic)

The term 'pesticides' or 'plant protection products' includes substances of very diverse chemical nature, most of which are used in agriculture against parasites and enemies of crops. Agriculture, however, uses only part of the production since these products are also widely used by industries (protection of textiles and wood,...), individuals, and even in

Table 10.9 *Annual increases (mg kg^{-1} dry soil per year) of cadmium concentrations in the 0-25 cm horizon of plots receiving various fertiliser materials since 1929 on the INRA estate in Versailles (after Tauzin and Juste, 1986)*

	From 1929 to 1984
Without a spread	0.0007*/0.0016**
Ammonium phosphate	0.0023
Superphosphate of lime	0.0032
Natural phosphates	0.0025

*From 1929 to 1960; **from 1960 to 1984

rural areas for non-agricultural purposes (weeding of roads and railway tracks, anti-mosquito measures...). It is necessary to bring relativism into the concept of soil pollution by pesticides to the extent that their toxicity is desirable. Their effects on the soil and vegetation must be taken into account as well as their subsequent transport to groundwaters and rivers.

Since 1885, use of copper salts (Bordeaux or Burgundy mixtures) against a fungus, mildew, has been responsible for the accumulation of this metal in all vineyard soils. The amounts used varied from 15 to 40 kg copper per hectare and per year until the availability of organic or organometal bactericides, i.e., for three-quarters of a century. It is therefore not surprising that many vineyard soils now contain more than 500 mg copper per kg soil. Copper is strongly fixed at the surface and migrates to a shallow depth. The relative resistance of an existing vine to copper is explained by the fact that its roots are below the zone of accumulation. Accidents occur after an old vine is uprooted, when another vine or a different plant replaces it. The excess copper results in considerable reduction in growth of the new plant and may cause its death. The toxicity of this metal is manifest mainly in acid soils. A supply of lime, reducing soil acidity, may mitigate the effects of copper toxicity (Delas and Juste, 1975).

As opposed to mineral pesticides, organic pesticides are characterised by considerable diversity of the molecules that compose them. In France, about 450 active homologous materials found in 3500 commercial products were surveyed. As soon as they arrive on the soil, organic pesticides are subjected to a fairly rapid microbial degradation accompanied by the appearance of metabolites whose properties and behaviour in the soil differ markedly from those of the parent molecule. It may be noted that studies of most organic pesticides have shown formation in the soil of 'bound residues', which are metabolites that cannot be extracted with the solvents available. It is not possible to specify the chemical identity or subsequent effects of these bound residues on the environment (Barrusio et al., 1991; C. Costé, pers. comm.; Soulas, 1991).

According to a comprehensive study published in 1990, DDT, banned in agriculture since 1972 but highly residual like all chlororganic products, is still present in rivers. Lindane, used mainly on maize and rape, is the last of the chlororganic substances used in large quantities. It is one of the principal water pollutants. In the major French estuaries, its concentration in the water is about 0.005 to 0.01 µg L^{-1}. The average detected content in waters of rural basins varies from 0.02 µg L^{-1} in Brittany to 0.3 µg L^{-1} in Champagne and 1 µg L^{-1} in the Northern Province. Contamination of groundwater varies from region to region, the values ranging from 0.002 µg L^{-1} to 3 µg L^{-1}. The triazines, such as atrazine used mainly for maize, and simazine, used for maize and grapevines, may persist for several months in the soil and water. Atrazine contents of 0.8 µg L^{-1} have been recorded in the Loiret aquifer. Concentrations reaching 1.3 µg L^{-1} for atrazine and 0.14 µg L^{-1} for simazine have been detected in Charente-Maritime (Bélamie and Giroud, 1990). In comparison, the maximum permissible concentration of pesticides in waters intended for human consumption is 0.1 µg L^{-1} per substance and 0.5 µg L^{-1} in total for all substances (European Directive 80/778/EEC and decree of 3 January 1989). With improved analytical methods, measurements obtained since 1990 in the framework of systematic surveillance of both surface and groundwaters reveal that the variety of pesticidal substances in them is more than hitherto believed. It is still difficult to obtain comprehensive data for France. The problem of water contamination by pesticides is of great concern and the relevant administrations are making efforts to curtail it (R. Bélamie, pers. comm.).

Recent studies indicate that the organic pesticides presently in use may disturb microbial activity in soils for several weeks and sometimes several months after treatment but, on the other hand, they do not appear to have an effect beyond this period when the products are applied in amounts recommended by the manufacturers (Cernakova et al., 1991; C. Costé, pers. comm.; Junnila et al., 1993; Lafrance et al., 1992; Olson and Lindwall, 1991; Tu, 1994a, 1994b; Wardle and Parkinson, 1992). Additional investigations are necessary to evaluate the risk when monocultures need frequent applications of herbicides (Perucci and Scarponi, 1994).

The toxicity of pesticides for earthworms is extremely variable. Herbicides such as atrazine, simazine or 2.4-D do not appear to have a significant effect when used in normal quantities. Concomitantly, fungicides such as benomyl or even insecticides-cum-nematicides such as ethoprop are highly toxic: reduction in biomass of earthworms by 95% and 88% respectively three weeks after treatment (Lee, 1985; Potter et al., 1990; Potter et al., 1994).

With Irrigation Waters

Irrigation, which has greatly increased over 40 years (see Table 10) and, more specifically, poor water management of irrigated areas are the main factors in soil salinisation. It has been estimated that every year 10 Mha are rendered useless for farming because of this phenomenon (compared to 3031 million cultivated hectares).

Table 10.10   *Increase in irrigated areas in the world since 1800 (after Szabolcs, 1994)*

Year	Irrigated areas (Mha)
1800	8
1900	48
1949	92
1959	149
1980	230
1990	265

Salinisation is an excessive accumulation of highly soluble salts (e.g. chlorides, sulphates, carbonates of sodium and magnesium) in the surface layers of soils. The extremely high osmotic pressure of the soil in the liquid phase affects the water supply to plants. Furthermore, some ions may have a toxic effect (Na, Cl, B, Se). Lastly, after fixation on clays, sodium may degrade the physical properties of the soil (impermeability).

Salinisation may be due to the actual quality of irrigation waters with a too high salt content. Otherwise, often excessive supplies of water are responsible for a rise in the water table. The mechanisms of capillary rise, unimportant so long as the water table is deep, become extremely active and the salts dissolved in the water table, even if in low concentration, precipitate in that part of the soil colonised by the roots. These phenomena are very important when climatic conditions promote evapotranspiration (e.g. arid zones) and when drainage is inadequate (C. Cheverry and G. Bourrié, pers. comm.; Robert, 1992; Szabolcs, 1994).

Within 'Polluted Sites'

'Polluted sites', earlier termed 'black spots', are sites in which the soil or subsoil or groundwaters have been polluted locally by former dumps of wastes or by infiltration of polluting substances. These are concentrated pollutions that generally justify anti-pollution treatments. The national survey carried out in 1994 by the Ministry of Environment surveyed 669 'polluted sites' (Table 10.11). The survey was not exhaustive and dealt only with sites known to the administration in which the soils are polluted or are responsible for a pollution observed in the surface or groundwaters. The size of the surveyed 'polluted sites' varies from a few square metres to several hectares.

Table 10 11   *Principal activities responsible for pollution observed during the 1994 survey (source: Ministry of Environment, 1994)*

Activity	Polluted sites (%)
Chemical, parachemical, pharmaceutical industries	19.5
Ferrous metal industry	18.0
Treatment and elimination of wastes	14.3
Oil and natural gas industry	10.6
Cokeries and town gasworks	9.6
Non-ferrous metal industry	7.0
Others	21.0

There are three major categories of polluted sites:

- Old dumps with no regard for present technical regulations, notably dumps on fragile subsoils and for which a pollution of groundwaters has been recorded (255 sites).

When the sites were in use, technical and scientific knowledge about the nuisances and risks caused by dumping wastes was limited or even non-existent, and sensitivity to environmental problems negligible.

- Dumps of wastes or abandoned chemical products following failure of companies or due to fraudulent practices while introducing or eliminating the wastes (129 sites).

- Soils polluted by fallouts, infiltrations or dumping of polluting substances associated with the past or present use by an industrial installation or a transport accident (357 sites).

Some industrial sites, abandoned or active, may comprise several sources of pollution. It may be noted that less than 4% of recorded pollutions are of accidental origin (Ministry of Environment, 1994) (see Tables 10.11 and 10.12).

Table 10 12 *Nature of pollutants in soil and groundwaters at polluted sites surveyed in 1994*

Pollutants	No. of polluted sites
Arsenic	47
Barium	14
Cadmium	38
Cobalt	7
Chromium	82
Copper	59
Mercury	30
Molybdenum	3
Nickel	34
Lead	100
Selenium	2
Zinc	85
Cyanide	48
Hydrocarbons	231
PAH*	66
PCB** PCT***	34
Halogen solvents	62
Non-halogen solvents	35
Pesticides	18
Others	238

*Polycyclic aromatic hydrocarbons **Polychlorobiphenyls ***Polychlorotriphenyls
(*Source:* Ministry of Environment, 1994)

CASE OF NITRATES: BEWARE OF FALSE IDEAS!

Soils are not polluted by supplies of nitrogen. In fact, soils naturally contain in their arable layer about 5 to 10 tons nitrogen per hectare, most of which is incorporated in organic matter. Annual supplies of nitrogen in the form of mineral fertilisers or effluents from animal farms rarely exceed 500 kg per hectare, which is less than 10% of the total nitrogen reserve and there are no phenomena of accumulation over several years, as is the case for metals.

However, when nitrogen in the soil appears in the form of nitrates in surface and groundwaters, there is risk of water pollution. To understand this, it is necessary to compare the quantity of mineral nitrogen present at any time in the soil with the dynamics of water (infiltration and runoff) and with the extraction of nitrogen by vegetation. The mineral nitrogen present in cultivated soils is certainly sourced from the supply by man and also originated from mineralisation of organic matter in the soil. This mineralisation may be considerably increased by some agricultural practices, for example tilling grasslands (Sébillotte and Meynard, 1990). An increase in nitrate content in groundwaters is a common occurrence in France and is highly variable from place to place. Thus, in aquifers on the alluvial terraces along the Garonne river, during summer the nitrate content in waters varies from 0 to 278 mg L^{-1} with an average of 60 mg L^{-1} (Fustec et al., 1991). In comparison, the maximum permissible concentration in water for human consumption is 50 mg L^{-1} (European directive 80/778/EEC of 15 July 1980).

Nitrates are the principal source of nitrogen for plants but an excessive accumulation of them in harvests is another aspect of nitrate pollution. The quantity of nitrates present in a plant at any given moment depends on the quantity available in the soil as well as such diverse factors as the nature of the plant and intensity of light, since light plays a very important role in plant metabolism. Thus a reduction of 28% nitrate content was observed in spinach after 12 hours of illumination (Blanc and Morisot, 1980). A study conducted on about 400 meals throughout France indicated the importance of solid foods, mainly vegetables, in the daily amount of nitrates and nitrites ingested by French people. The actual supply from drinking water is only 22% of the total for the entire population and one-third for children. The nitrate content in highly 'concentrating' vegetables (beetroot, radish, marrow, spinach, ...) is as much as 2000 to 4000 mg kg^{-1}. In less 'concentrating' vegetables, but consumed in large quantity, such as potatoes, cabbage and carrots, the concentrations have greatly increased in recent years and contribute 45% of the total supply in meals. It was observed that in 80% of the samples of canned baby foods analysed, the contents were more than 50 mg kg^{-1}, the prescribed standard for solid food for an infant of less than three months (amended ministerial order of 1 July 1976 on dietetic foods and infancy) (Ministry of Social Affairs and Integration, 1992).

AND THE NATURAL PEDOGEOCHEMICAL BASE?

Trace-metal elements are found naturally in soils, inherited from the parent rock as well as from atmospheric fallouts of natural origin. These elements are absorbed partly by plant roots and later released either in the soil (by decomposition of roots) or at the surface (by debris of aerial parts, forest litter). Two questions now arise: 1) What is the relationship between the natural contents of trace-metal elements in soils and other soil properties, in particular the clay and iron contents? and 2) How to distinguish between contamination of anthropic origin and natural contents? (Baize, 1994, 1997). Table 10.13 shows that the natural 'pedogeochemical' base is extremely variable from one soil to another.

Soils, like all components of the biosphere, also contain radioactive elements and isotopes of natural origin. Of these, three that are always present and with a very long half-life originate from the parent rock: uranium 238 (half-life 4.5×10^9 years; contents between 10 and 120 Bq kg^{-1}), thorium 232 (half-life 1.4×10^{10} years; contents between 10 and 120 Bq kg^{-1}) and potassium 40 (half-life 1.3×10^9 years; contents between 100 and 200 Bq kg^{-1}). Uranium 238 and thorium 232, parents of radioactive families, spontaneously produce other natural radioisotopes, the principal ones being protactinium, actinium, radium, bismuth and lead. If these daughter elements are in equilibrium with their parent, the activity of each (in Bq kg^{-1}) is equal to that of the parent. Nonetheless, under certain soil conditions, this equilibrium may be disturbed and one or several daughter elements may disappear completely from the soil.

Table 10.13 *Average values of contents of trace-metal elements (expressed in mg kg^{-1}) and of clay and iron (expressed in g 100 g) for several soils in northern France considered free from contamination, or very slightly contaminated by human activity*

		Clay	Fe	Cd	Cr	Cu	Ni	Pb	Zn
I.	e	17	1.8	0.22	44	11	18	21	53
	b	27	2.9	0.17	62	13	30	18	67
II.	e	15	1.2	0.08	37	6	11	23	34
	b	35	3.1	0.04	64	11	26	26	55
III.	a	38	3.6	0.34	70	12	37	43	101
	b	58	4.8	0.17	96	13	51	35	114
IV.	e	32	4.2	0.43	86	19	59	107	237
	b	51	7.4	0.75	141	33	162	101	437

I. Deep silty saline soils of the French Vexin (Neoluvisols formed from silty loess): e—upper horizons; b—deeper horizons.

II. Deep silty soils, highly hydromorphic, in the south-eastern part of the Paris basin (Yonne, Seine-et-Marne) (degraded Luvisols formed from old loess): e—upper loess horizons; b—deeper clay-loess horizons.

III. Clay soils, reddish in colour, non-calcareous on the plateaus of Burgundy (Yonne, Côte d'Or) ('Red dawns'): a—upper humiferous clay-loess horizons; b—deeper clay horizons.

IV. Clay soils, thick and with a high iron content on the Sinémurian platform north and east of Morvan (Yonne, Côte d'Or): e—upper loess-clay horizons; b—deeper clay horizons (source: D. Baize, pers. comm.; D. Baize, 1997).

Further, beryllium 7 (half-life 58 days; contents generally less than 20 Bq kg^{-1}) is constantly formed in the upper layers of the atmosphere due to the effect of cosmic rays (source: IPNS, pers. comm.).

Glossary

Pollution
: (*etymologically, to pollute signifies to profane, to soil, to render dirty or to degrade*). Pollution is thus an unfavourable change in the natural environment which appears entirely or partly as a by-product of human activity through direct or indirect effects that alter the criteria of distribution in energy flows, radiation levels, physicochemical composition of the natural environment and abundance of living species. These modifications may affect human beings directly or through agricultural resources, water and other biological products. They may also affect them by changes in the physical objects they possess, the recreational possibilities of the environment or even destruction of natural beauty (Ramade, 1989).

Soil
: The upper layer of the earth's crust composed of mineral particles, organic matter, water and organisms (definition proposed by the International Organisation for Standardisation).

Traces (in a state of traces, trace elements)
: Said of a substance present in any environment in concentrations of less than 1 g kg^{-1}.

⌐ URTHER READING

Baize D. 1994. Teneurs totales en métaux lourds dans les sols français, premiers résultats du programme ASPITET. Courrier de l'Environnement de INRA, no. 22, juin 1994, pp. 37-45.

Baize D. 1997. Teneurs totales en éléments traces métalliques dans les sols (France). Références et stratégies d'interprétation. INRA ed., Paris, 410 pp.

Barrusio E, Andrieux F, Schiavon M, Portal JM. 1991. Intérêts et limitations des méthodes de séparation des micropolluants organiques des sols. Sci. du Sol, 29 (4): 301-320.

Bélamie R, Giroud S. 1990. Les pollutions liées à l'utilisation des pesticides. Perspectives Agricoles, no. 146, avril 1990, pp. 52-56.

Blanc D, Morisot A. 1980. Les nitrates d'origine agricole: leur accumulation dans la plante, leur effet sur l'environnement. Ann. Nutr. Allm., 34: 791-806.

Bonneau M. 1992. Pollution atmosphérique et dépérissement des forêts dans les montagnes françaises, synthèse des recherches. In: Programme DEFORPA, rapport 1992. Ministère de l'Environnement, pp. 7-34.

Bourrié G. 1993. Parcelle de Fougères (Ille-et-Vilaine, région Bretagne); rapport sur les teneurs en radio-éléments 1986. Observatoire de la Qualité des Sols. Ministère de l'Environnement.

Boutron C. 1998. Le plomb dans l'atmosphère. La Recherche, 19 (198): 446-455.

Cernakova M, Kurukova M, Fuchsova D. 1991. Effect of the herbicide Bentanex on soil microorganisms and their activity. Folia Microbiol. 36 (6): 561-566.

Coppenet M, Golven J, Simon JC, Le Corre L, Le Roy. 1993. Evolution chimique des sols en exploitations d'élevage intensif: exemple du Finistère. Agronomie (1993) 13: 77-83.

Delas J, Juste C. 1975. Quelques problèmes posés par les sols viticoles acides. Connaissance de la vigne et du vin, 2: 67-80.

Dumont JC, Vernette G. 1974. Observations relatives au plomb échangeable des sols prélevés en bordure de la route nationale 10 dans la région des Landes. Bull. Inst. Géol. Bassin Aquitanine, 16: 121-126.

Fustec E, Schenck C, Cloots-Hirsch, Soulié, M, Bouton D. 1991. Les nitrates dans les vallées fluviales. Programme Environnement du CNRS, 51 pp.

Gaber J. 1994. La contamination des sols le long des routes, une approche de la pollution atmosphérique. *Biological and technical methods of activities against the negative effect of motorization in the environment.* Materialy, Sympozjium Naukowego—czesc II, Cracovie, 8 au 10 juin 1994. SETRA, Ministère de l'Equipement, pp. 25-38.

Godin PM. 1985. Modelling of soil contamination by airborne lead and cadmium around several emission sources. Environmental Pollution (Series B) 10 (1985): 97-114.

Harrison RM, Laxen DPH, Wilson SJ. 1981. Chemical associations of lead, cadmium, copper, and zinc in street dusts and roadside soils. Environ. Sci & Technology, 15 (11): 1378-1383.

Hutton M. 1983. The environmental implications of cadmium in phosphate fertilizers. Phosphorus and Potassium, 123: 33-36.

Junnila S, Heinnoken-Tanski H, Erivö LR, Laiten P. 1993. Phytotoxic persistence and microbiological effects of metribuzin in different soils. Weed Research, 33: 213-223.

Knap AH (ed.). 1990. The Long-Range Atmospheric Transport of Natural and Contaminant Substances. Kluwer Academic Publishers, Dordrecht, the Netherlands, pp. 197-229.

Lafrance P, Salvano E, Villeneuve JP. 1992. Effet de l'herbicide atrazine sur la respiration et l'ammonification de l'azote organique dans un sol agricole au cours d'une incubation. Can. J. Soil Sci., 72: 1-12.

Laxen DPH, Harrison RM. 1977. The highway as a source of water pollution: an appraisal with the heavy metal lead. Water Research, 11: 1-11.

Lebreton L, Thévenot DR. 1992. Pollution métallique relargable par les aérosols d'origine autoroutière. Environ. Technology, 13: 36-44.

Lee KE. 1985. Earthworms. Their Ecology and Relationships with Soils and Land Use. Academic Press, North Ryde, New South Wales, Australia, 411 pp.

Leprêtre A. 1995. Etude statistique des paramètres physico-chimiques et répartition des métaux lourds du site expérimental d'Evin-Malmaison. Observatoire de la Qualité des Sols. Ministère de l'Environnement.

Leprêtre A, Dessaint F. 1988. Etude statistique des paramètres physico-chimiques et répartition des métaux lourds du site expérimental de Neuf-Berquin. Observatoire de la Qualité des Sols. Ministère de l'Environnement, 37 pp.

Martin S, Berteau I, Vassiliadis A. 1993. Le curage des cours d'eau lorsque les sédiments contiennent des éléments-traces toxiques. Le Courrier de l'Environnement de l'INRA, 20: 27-35.

Minstère de l'Environnement. 1988. La Valorisation Agricole des Boues de Stations d'Epuration Urbaines. Cahier Technique de la Direction de l'Eau et de la Prévention des Pollutions et des Risques, no. 23, 117 pp.

Ministère de l'Environnement. 1992. Rapport au Président de la Commission des Communautés Européennes établi en application de l'article 17 de la Directive 17 du Conseil du 12 juin 1986 relative à la protection de l'environnement et notamment des sols, lors de l'utilisation des boues d'épuration en agriculture par la Direction de l'Eau en décembre 1992.

Ministère de l'Environnement. 1994. Recencement des sites et sols pollués 1994 (Direction de la Prévention des Pollutions, Service de l'Environnement Industriel), 277 pp.

Ministère des Affaires Sociales et de l'Intégration. 1992. La diagonale des nitrates. Etudes sur la teneur en nitrates de l'alimentation, 20 pp.

Nriagu JO. 1989. A global assessment of natural sources of atmospheric trace metals. Nature, 333: 47-49.

Nriagu JO, Pacyna JM. 1988. Quantitative assessment of worldwide contamination of air and soils by trace elements. Nature, 338: 134-139.

Olson BM, Lindwall C-W. 1991. Soil, microbial activity under chemical fallow conditions: Effects of 2,4-D and glypholate. Soil Biol. Biochem., 23 (11): 1071-1075.

Perucci P, Scarponi L. 1994. Effects of the herbicide imazethapyr on soil microbial biomass and various soil enzyme activities. Biol Fertil. Soils, 17: 237-240.

Potter DA, Buxton MC, Redmond CT, Patterson CG, Powell AJ. 1990. Toxicity of pesticides to earthworms (Oligochaeta: Lumbricidae) and effects on thatch degradation in Kentucky bluegrass turf. J. Econ. Entomol., 83 (6): 2362-2369.

Potter DA, Spicer PG, Redmond CT, Powell AJ. 1994. Toxicity of pesticides to earthworms in Kentucky bluegrass turf. Bull. Environ. Contam. Toxicol., 52: 176-181.

Puxbaum H. 1991. Metal compounds in the atmosphere. In: Metals and Their Compounds in the Environment. E. Merian (ed.). VCH, Weiheim (All.), pp. 257-286.

Ramade F. 1989. Eléments d'écologie, écologie appliquée. Masson éd., Paris, 452 pp.

Robert M. 1992. Le sol, ressource naturelle à préserver pour la production et l'environnement. Cahiers Agricultures, 1992-1, pp. 20-34.

Saure E, Juste C. 1994. Enrichment of trace elements from long-range aerosol transport in sandy podozolic soils of southwest France. Water, Air and Soil Pollution, 73: 235-246.

Schlesinger WH. 1991. Biochemistry: an Analysis of Global Change. Acad. Press, San Diego, CA, 443 pp.

Sébillotte M, Meynard JM. 1990. Systèmes de culture, systèmes d'élevage et pollutions azotées. Nitrates, agriculture, eau. Paris, 7-8 novembre 1990. INRA Editions, Paris, pp. 289-312.

Six P. 1992. La qualité de l'eau et des sédiments en cause... Quelques accidents spectaculaires. "La restauration des rivières et des canaux", Colloque de Bouvines, sept. 1992. Agence de l'Eau. Artois-Picardie.

Soulas G. 1991. La biodégradabilité des pesticides dans les sols. C.r. du 21[ieme] Congrès du Groupe Français des Pesticides, 22-23 mai 1991, Dijon, pp. 121-137.

Tauzin J, Juste C. 1986. Effet de l'application—à long terme de diverses matières fertilisantes sur l'enrichissement en métaux lourds de parcelles nues. Conv. d'étude Ministère de l'Environnement/INRA, 4193: 76 pp.

Tu CM. 1994a. Effects of herbicides and fumigants on microbial activities in soil. Bull. Environment. Contam. Toxicol., 53: 12-17.

Tu CM. 1994b. Effects of some insecticides on microbial activities in sandy soil. J. Environ. Sci. Health. B(29)2: 281-292.

Wardle DA, Parkinson D. 1992. Influence of the herbicides 2,4-D and glyphosate on soil microbial biomass and acitvity: a field experiment. Soil Biol. Biochemical., 24(2): 185-186.

Wiart J, Verdier M. 1995. Etude de la teneur en éléments-traces métalliques des boues de stations d'épurations urbaines françaises. Rapport AGHTM & FNDAE, 60 pp.

Wopereis MC, Gascuel-Odoux C, Bourrié G, Soignet G. 1988. Spatial variability of heavy metals in soil on a one-hectare scale. Soil Science, 146 (2): 113-118.

Yassoglou N, Kosmas C, Asimakopoulos J, Kallianou C. 1987. Heavy metals contamination of roadside soils in the Greater Athens area. Environ. Poll., 47: 293-304.

CHAPTER 11

Contamination of Soils by Heavy Metals and Other Trace Elements

P. Cambier, M. Mench

INTRODUCTION AND DEFINITIONS

In several scientific disciplines, chemical elements are conventionally separated as major elements and 'traces' based on thresholds of concentration. The word 'traces' evokes analytical difficulty in determining low concentrations, with no relationship to any property or function. Trace elements can therefore be metals (e.g.: Cd, Cs, Cu, Ni, Pb, Zn, ...) or non-metals (e.g.: F, Cl, B, I, ...). Metals are defined according to their physical properties (high electrical and thermal conductivity, high reflectivity, malleability), which are largely explained by their chemical properties (essentially the tendency to donate electrons).

The phrase *heavy metals* is often wrongly used to describe trace elements. Heavy metals were originally defined according to their properties for combination with sulphides which yield dense solids with limited solubility. It may now be considered that they are metal elements with high atomic weights, reacting in cationic form at the level of 'soft' polarisable electronic orbitals. Their usages are based on their properties in electricity, electronics, for the treatment of metal surfaces and for reinforcement of plastic materials. Because of their toxicity, they are used for crop protection (pesticides), in paints for boat hulls etc. These properties of toxicity thus explain the negative connotation of the phrase 'heavy metal'.

> With respect to heavy metals or trace elements, the first point to emphasise is that, contrary to synthesised organic molecules which can be degraded, these elements exist naturally and are permanent, playing a role in the biogeochemical cycles. They represent a component of soil quality because they participate in significant soil functions: ecosystem, substrate of plant production and filter for entrance to aquifers.

TRACE ELEMENTS IN OUR ENVIRONMENT

Geochemist's Point of View

In Earth Sciences, trace elements are those present in the lithosphere in concentrations of less than 0.1%, comprising 68 of the 80 elements constituting the earth's crust. Only those elements that participate most in biogeochemical cycles are discussed here, without special consideration of the case of radioactive elements.

Soil geochemical background

Soils form the interface between the lithosphere, atmosphere and biosphere. In regions where there is little anthropic activity, their trace-element contents reflect those of a variable geochemical background. Due to the effects of climate, vegetation, topography and living organisms in the soil, the elements of parent rocks are redistributed between horizons, the different layers of the soil. This results in a pedogeochemical background with lateral and vertical variations (Baize, 1997).

Table 11.1 gives the values corresponding to the first and last deciles in the distribution of total contents of trace elements in the surface horizons of non-contaminated agricultural soils taken from various countries. Variations are often of the order 1 to 10. Minimum and maximum variations are much more but lack statistical significance. There are numerous correlations between the various properties of soils, in particular a positive correlation between the contents of trace elements and the clay or iron contents. The nature of the rock, however, also has a

Table 11.1   *Range (1st and 9th deciles) of concentrations of trace elements in the surface layers of agricultural soils considered non-contaminated in samples taken from various countries**

Data on the 'pedogeochemical' background			
Element	*France* (86 soils)	*Poland* (30-127 soils)	*Switzerland* (370-880 soils)	*USA* (3045 soils)
As	3.4-37	1.0-7.5		
Cd	0.07-0.51	0.1-1.4	0.19-0.5	0.05-0.56
Co	2.1-27	1.2-8.3	6.1-12.4	
Cr	22-102	4.7-59	21-49	
Cu	7-87	2.4-15.1	15-162	5.3-62
F			302-569	
Hg	0.02-0.15		0.052-0.24	
Mo			0.03-0.58	
Ni	7.51	2.5-22	19-50	5.6-40
Pb	19-62	8.3-43	17-35	5.0-20
Se		0.08-0.3		
Tl			0.027-0.12	
Zn	31-153	13.1-124	44-92	12.7-105

*France: total contents after dissolution with $HF + HClO_4$ (mg/kg soil)—Soil Analysis Laboratory, INRA, Arras.
Poland, USA: see references in Cambier, 1994.
Switzerland: contents after extractions in acid, 2M HNO_3 (Meyer, 1991).

decisive effect. Thus regions are marked by high contents of nickel or arsenic, cadmium etc.

Physicochemical reactions in soils

If interested not only in the presence but also the speciation and dynamics of trace elements in soils, they can be classified according to chemical properties, mainly their reactivity to oxygen and their solubility, depending on the pH (Buffle, 1988; Cambier, 1994).

The elements Cd, Co, Cu, Mn, Ni, Pb, Tl and Zn best meet the definition of heavy metals. They react as bivalent cations (except for thallium Tl^+), more strongly with soft electron donors such as the sulphide S^{2-} ion and some forms of phosphorus and nitrogen, than with oxygen or water. Nevertheless, in natural superficial environments such as soils, they often form oxides and hydroxides, carbonates, organic complexes with ligand sites such as COO^- as well as surface complexes with FeO^- or AlO^- of major metal oxides.

The principal chemical law that governs the dynamics of metals in this group is that their solubility and consequently mobility reduces as the pH increases. Actually, oxides and carbonates are more soluble in an acid environment and the complexes formed with ferric oxides or insoluble humic acids more stable at a basic pH.

Nevertheless, soils also contain soluble organic ligands, which are more dissociated from solid particles when the pH is basic. Therefore, in some situations in which organic matter plays a major role, a slightly basic pH may help in dissolution of elements, in particular Cu or Pb, which present a high affinity for these ligands.

In strongly reducing conditions, heavy metals can be immobilised as insoluble sulphides. In ordinary conditions of soils, Cd, Ni, Tl and Zn are the most mobile elements of this first group.

The trivalent elements, Bi, Cr, In, Sc, Tl and divalent mercury Hg(II), yield rather 'hard' acid cations. They combine effectively with O^{2-} and OH^- ligands. They notably form low solubility hydroxides at a neutral or slightly basic pH. The case of thallium illustrates the possibility for certain metals to present different valences in soils. Mercury may have valences 0, +1, +2. In the case of this element, what is noteworthy is the importance of its chemical speciation. Combined with carbon in a covalent bond, Hg forms organometal complexes that are stable and volatile. Its dynamics therefore involves all three phases of soils (gas, water, solid).

Elements As, Mo, Sb, Se, V and Cr in their hexavalent form, have high valences in soils as a result of which they have the properties of very 'hard' cations. They finally form highly stable anionic complexes with several oxygens: arsenate AsO_4^{3-}, chromate CrO_4^{2-}, selenate ions SeO_4^{2-}, selenite ions SeO_3^{2-} etc. Due to this negative charge their mobility increases with pH. At a neutral or slightly acid pH, these anions may be retained at sites on the surfaces of iron and aluminium oxides.

The role of organic matter is not well known for these elements and the effect of redox conditions is ambivalent: reducing conditions are favourable for species that are more strongly retained by certain constituents of soils (for example, $Cr^{3+} > CrO_4^{2-}$, $SeO_3^{2-} > SeO_4^{2-}$); but these constituents may be transformed, or even dissolved, in highly reducing conditions. This applies in particular to ferric oxides which are traps for almost all trace elements, and may be dissolved in the form of ferrous iron.

Point of View of a Biologist: Essential or Toxic Elements?

In Life Sciences, trace elements are defined as elements at concentrations lower than 100 mg kg^{-1} dry matter (DM) in biological organisms, plants or animals, not affected by deficiencies or toxicities (Coïc and Coppenet, 1989; Loué, 1993; Juste et al., 1994). They may be grouped in two classes:

- *essential*, because of their essential role in the structure and metabolism;
- *non-essential*, if they never display a beneficial effect and have no biological utility known to date.

Essential trace elements are also called *micronutrients*. However, as soon as their concentration in an organism exceeds a certain threshold, they can also become toxic. Table 11.2 provides a non-exhaustive list of trace elements, with an indication of their functions and impacts in relation to living creatures.

Considering more particularly soil ecosystems, excessive concentrations of trace elements induce dysfunctions of a biological type. Mineralisation of organic matter may be slowed down because of the toxicity of trace elements for micro-organisms (Ross, 1994; Juste and Mench, 1992). Some transformations of nitrogen may be inhibited. For example, Giller et al. (1993) have shown that the fixation of nitrogen by rhizobia is reduced in the presence of such metals as cadmium. The latter effects may be due to response of plants to toxicity (Ross, 1994).

Agronomist's Point of View: Soil-Plant Transfers and Plant Toxicity

For use in agriculture, a soil should ensure a satisfactory yield of quality crops whose composition ought never to be harmful for animals or humans. The sublethal plant toxicity due to trace elements is expressed by changes in metabolism (e.g.: synthesis of plant chelates). Severe plant toxicity is manifested by symptoms of chlorosis, reduction in aerial biomass, or a limited development of roots that always signify harvest losses. From the viewpoint of transfer of elements, plants and water are the principal routes between soils and other ecosystems. In the absence of a significant supply to the leaves, there is a relation between the total content of trace elements in the soil and their concentrations in the soil solution and the plants. This relation depends, however, on many factors associated with the plant and the soil.

Table 11 2   *Importance of trace elements in relation to living organisms*

	Essential or beneficial for plants	Essential for animals	Toxic for plants	Toxic for animals
As	No	Yes	Yes	Yes
B	Yes	No	Yes (narrow range between utility and toxicity)	
Be	No	No	Yes	Yes
Cd	No	No	Yes	Yes
Cl	Yes	Yes		
Co	Yes (leguminous plants)	Yes	Yes, at high content	Yes
Cr	No	Yes	Yes, especially Cr(VI)	Yes
Cu	Yes	Yes	Yes	
F	No	Yes	Yes	Yes
Fe	Yes	Yes		
I	No	Yes		Yes
Hg	No	No		Yes
Li	No	Yes		
Mn	Yes	Yes	Yes (especially at pH < 5 and high content)	
Mo	Yes	Yes		Yes (narrow range)
Ni	Yes	Yes	Yes	Yes
Pb	No	No	Yes	Yes
Se	Yes	Yes	Yes	Yes (narrow range)
Sn	No	Yes		Yes
Tl	No	No		Yes
V	Yes	Yes	Yes	Yes (narrow range)
Zn	Yes	Yes	Yes, at high content	

In conclusion, the dynamics of trace elements in soils and their biological effects, especially on plants and their quality, are interdependent. It is not enough to consider the total contents of trace elements to assess the contamination of a soil. The natural contents are also highly variable. A classic approach in agronomy consists in estimating the bioavailability of the elements based on a chemical extraction from a soil sample. Extraction in non-buffered saline solutions (e.g.: $Ca(NO_3)_2$ or $CaCl_2$ 0.1N) generally provides values that correlate well with the availability of Cd, Co, Mn, Zn and Ni. For Cu, as also for Mn, Ni, Zn, satisfactory results have been obtained with chelating solutions at a neutral or slightly basic pH (e.g.: DTPA or EDTA). Nevertheless, there is no universal extractant and the interpretation must take into account factors associated with the soil and the plant (Coïc and Coppenet, 1989).

The main factors associated with a plant are the species and the variety, the stage of development, and the interactions between nutritive elements. On a particular soil the concentration of Cd may vary from 0.01 mg kg^{-1} DM in a barley grain to more than 0.5 mg kg^{-1} DM in leaves of lettuce and spinach, but this may range between 1.8 and 8 mg kg^{-1} DM depending on the particular varieties of lettuce. Localisation of elements in the organs varies depending on the species: maize accumulates 60% of the Cd absorbed in the roots while tobacco stores it mainly in the leaves (80%). The thresholds of toxicity also depend on the plant. Cabbage, turnip, beetroot and lettuce are sensitive crops while maize, soy bean and cereals are tolerant. Graminaceous forage plants are extremely tolerant, with toxicity of thresholds of 10 mg kg^{-1} DM for Cd and from 200 to 400 mg kg^{-1} DM for Zn.

Iron, zinc and manganese are generally the metals extracted most by plants. However, the availability of other trace elements in the substrate may modify this. The coefficient of transfer, obtained by computing the relation between the contents in a plant and the soil is between 0.5 and 10 for Cd and Zn, between 0.01 and 1 for Cu and Ni and between 0.01 and 0.5 for Pb. It is about 0.5 for Tl. Toxicity also depends on the chemical form of the particular element. Thus the elements As, Hg, Pb, Pd, Sn and Tl may remain in the form of organometal complexes in which the toxicity is much higher than that of inorganic forms.

Soil and climatic factors influence the composition of a soil solution (e.g.: pH, carbonates, state of the metal in solid phase) or the metabolism of a plant (e.g.: temperature of the roots, lighting, mineral nutrition). The general laws indicated in the earlier part of this chapter for the solubility of elements may in a first approximation be applicable to bioavailability.

For example, a nitrogenous fertiliser in the ammoniacal form leads to acidification of the soil, resulting in twice as much removal of Cd by rye-grass compared to removal with a nitrate fertiliser.

CONTAMINATION OF SOILS BY TRACE ELEMENTS

Sources: Diffuse and Local Contaminations

(also see Chapter 10 in this book, 'Sources of Soil Pollution')

So-called diffuse contamination is one which results in an increase in the content of a pollutant over time over a large area but which is not perceptible in year-to-year variation. Remote atmospheric fallouts contribute significantly to this type of contamination by the elements As, Cd, Hg, Mn, Ni, Pb, Sb, Se, Zn. The emissions are a result of natural processes (sand-storms, major forest fires, volcanic eruptions) and industrial and urban activities (burning coal, production of non-ferrous metals, use of leaded petrol, incineration of household wastes). The composition of fallouts due to the treatment of ores varies according to the precise composition of the latter. The iron and steel industry may be responsible for contaminations by arsenic, cadmium, fluorine etc. Ores of zinc and lead often have a high cadmium content. Productions of copper and gold cause contamination by arsenic. In fact, fallouts from these activities need to be taken into account at different scales (see below).

Other sources of diffuse contamination are fertiliser materials and crop-protection products. Fertilisers based on mineral deposits and organic amendments (liquid manures, ordinary manures, composts, sewage, sludges) provide significant quantities of trace elements. Fertilising with micronutrients such as Cu, Zn, Mn, B, Mo and Se introduces these elements into soils which, in principle, contain little of them. Many pesticide treatments, especially those based on copper, arsenic and mercury, have contributed or will contribute to contamination of soils.

Local contaminations are of very diverse origins (Ministry of Environment, 1994). Mining, industrial and urban activities may contaminate soils locally and seriously through the atmosphere or by liquid discharges and runoffs, and also through the atmosphere in a diffuse manner. For example, contamination by lead from exhaust gases is reduced exponentially near roads where there is a large volume of traffic, but may also be observed on a very large scale (Fig. 11.1).

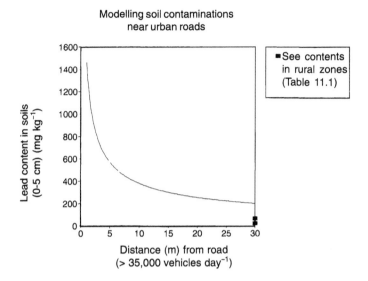

Fig. 11.1   Example of modelling variations of lead concentrations on the soil surface (depth 0-5 cm) near roads where there is heavy traffic (after Yassoglou et al., 1987).

Consequences (Examples)

The first part of this chapter dealt with the complexity of chemical and biological processes involving trace elements in the soil–water–plant system. A study of these should enable a better understanding and forecasting of the dynamics of these elements in both contaminated and non-contaminated sites. A few current problems as a result of contamination in soils by trace elements will now be analysed, starting with their consequences. In the last part of this chapter, we shall consider the measures to be adopted and the possible remedies.

Protection of food chain: urgent measures for cadmium

Diffuse contamination of soils by trace elements generally does not create serious problems. Nevertheless, the question of cadmium contents is central to many studies and measurements of the chemical qualities of soils. Among the heavy metals, Cd is relatively mobile and bioavailable. The joint expert committees of the FAO-WHO recommended that the weekly intake of cadmium should not exceed 7 µg kg^{-1} body weight, i.e., about 70 µg Cd per day per person. A normal diet, in various countries, supplies between 15 and 60 µg Cd per person per day, which leaves a small margin. A single example helps to understand the risks associated with contamination of soils: a lettuce plant grown in a market garden contaminated by 10 µg Cd kg^{-1} may have the same concentration in dry matter (a transfer factor of 1). As a result, a person consuming 50 g DM of this lettuce per week will ingest the tolerable weekly amount, taking into account what is provided by other foods (Godin, 1983).

In France, an estimation of daily provision of Cd in food indicated that the highest contributions are from cereal-based products, fresh vegetables and also meat, especially giblets. The contribution of cereals is also large in other European countries (22% in Germany, 26% in Great Britain) and that of potato as much as 20% in Germany. The concentrations of Cd in the consumed portions of fruits are low while they are high in spinach and mushrooms.

Local contamination by non-ferrous metal industries

Some areas near industrial sites, active or closed down, are contaminated by a number of metals, most often Cd, Cr, Cu, Pb, Zn, and as a consequence, soil uses are limited. Animal husbandry has had to be abandoned as also some crops (in market gardens), although this depends on the distance from the sources of pollution and the soil and climatic conditions. Thus in northern France, fallouts from a non-ferrous metal factory over an area of 10,000 hectares were estimated in 1989 to be 55 t for Pb, 44 t for Zn and 16.1 t for Cd. Contamination in forage plants affected cattle and now there are no animal farms. Photo 11.1 depicts part of an old industrial site in north-eastern Belgium, 'desert-like' because of an accumulation of Cd and Zn in the sandy soil (the grass-covered area represents an attempt at rehabilitation—see below). In this case, the subjacent water table was also contaminated by zinc and cadmium (Vansgronsveld et al., 1994).

There are other types of contamination, by non-metal trace elements. A known example in France is that of fluorine from iron and steel works in the Maurienne Valley. This essentially thionic element does not remain very long in soils and the damages observed were due mainly to fallouts on the aerial parts of plants. Nevertheless, contamination of soils leads to water pollution as well as less visible impacts on microorganisms. When arsenic release occurs, as in the neighbourhood of gold mines in Ghana, arsenate is partly retained in soils, with retention maximal at a pH of 4 to 5 aided by the presence of ferric and aluminium oxides. Although these conditions occur together in tropical soils, in this example from Ghana, the water and plants are seriously contaminated.

(a)

(b)

Photo 11 1   Abandoned site near an old zinc foundry in Belgium (a) and rehabilitation attempt by soil amendments and selection of resistant graminaceous plants (b) (Vansgronsveld, Limburgs Universiteit Centrum).

REMEDIES AND REGULATIONS

The example of cadmium has shown that it is important to maintain very low concentrations of certain trace elements in water, food, forage, tobacco and therefore in soils. This is ensured by good management of wastes and discharges of all types. For example, better control of discharges in waters has helped to reduce metal contents in sewage sludges. In the case of lead, the record of atmospheric fallouts in the Greenland ice cap (and in vintage wines! Lobinski et al., 1994) indicates that since the 1980s there has been a reduction associated with use of unleaded petrol. Questions with respect to treatment of polluted soils will now be discussed and also the regulations intended to limit the risks they represent.

Prospects of Rehabilitation of Sites Polluted by Trace Elements

Decontamination of soils by trace elements is not an easy matter, especially if the objective is actually to drastically reduce the total contents. The simplest method would be to assume the entire soil polluted and a waste to be removed. But the volume would soon become an insurmountable obstacle, given the fact that the possibility of dumping it in landfills is increasingly restricted and regulated: only 'ultimate' and 'inert' wastes may be deposited in dumps from the year 2002 in France.

It is possible to excavate soils polluted by metals and to treat them chemically (extraction by acid and reprecipitation) or by physico-chemical means (dispersion of the fine fractions and reflocculation). Sludges obtained, with a high concentration of contaminants, must be dumped in landfills and what remains of the soil is an altered material. The cost of such an operation poses a major constraint, however.

When contamination is extensive, it is presently only possible to think of rehabilitation operations to reduce the impacts or risks. They involve techniques of confinement, physical or physicochemical stabilisation or dilution (Logan, 1992).

For example, to attenuate the phytotoxicity sometimes observed in vineyard soils contaminated with copper, deep tillage can be done before planting new young plants. The soil is 'ripped', which results in mixing the surface horizon with lower layers, and the contamination is diluted. Other solutions are selection of a tolerant crop, liming to moderately raise the pH, or organic amendments to fix the copper. In Japan during the 1980s, contamination of soils by cadmium was urgently treated by covering rice fields with 25 to 30 cm of clean earth.

In the case of contamination by zinc and cadmium as shown in Photo 11.1, rehabilitation consisted of selecting and planting a tolerant vegetation and amending the soil with a material—'beringite', a mineral obtained from clay ash—which fixes metals. These operations limit infiltration and erosion and thereby spread of contamination, and eventually the land can be used for certain purposes.

The basic amendments, employed mainly for their effects on the physical properties of soils, also help to correct aluminium and manganese toxicities. This kind of treatment is not always sufficient for certain contaminations, however. For instance, during experiments conducted by INRA-Bordeaux on soils contaminated by Cd and Ni, only incorporation of iron or manganese could perceptibly reduce the mobility of these elements and the phenomena of plant toxicity (Photo 11.2). Further, cases of contamination by elements such as arsenic or mercury cannot be solved by raising the pH level because of their physicochemical behaviour.

Extraction of pollutants could be a more satisfactory option for the future. Decontamination by plants with recovery of metals from plant ashes after harvest is under consideration. This would involve using species that can absorb large amounts of metals (metallophytes) and carrying out genetic modification to achieve plants with developed aerial parts and concentrated metals. These decontaminations might take several years but could be adopted over vast areas, creating fewer economic and ecological problems than chemical treatments would. After all, some present contaminations originated at the beginning of the industrial era; the final solution to decontaminations merits a little perseverance.

Photo 11.2 Experiments to suppress plant toxicity on tobacco: (left) soil contaminated with sludges containing cadmium and nickel; (right) same soil amended with an iron powder. Sappin-Didier, INRA-Bordeaux.

Regulations

The only French regulations applicable to trace elements in soils concern spreading of sludges from waste water treatment: it is forbidden to spread sludges on agricultural soils in which the content of one or several elements is, or would be greater after the spread than the maximum value fixed by the norm NF U44-041 (Table 11.3). Furthermore, the pH of the soil should be more than 6. Other constraints are based on the moisture content in the field, type of agricultural production, proximity of water points and habitation, and the chemical and sanitary qualities of the sludges.

Table 11.3   *Maximal concentrations beyond which spread of sludges from urban waste water stations is prohibited*

Element	NF U44-041 (France, 1985)	EEC directives 1986-1988 (higher values*)
Cd	2	3
Cr	150	200
Cu	100	140
Hg	1	1.5
Ni	50	75
Pb	100	300
Se	10	—
Zn	300	300

*The EEC gives value ranges.

Apart from these spreads, the protection of soils is looked after by the prefect and delegated services (DAAS, DRIRE, ...). Some countries have further improved their regulations, for example the Netherlands, which has stipulated maximum values that vary depending on soil texture (clay and organic matter content), and Switzerland, which takes into account not only the total contents, but also the 'soluble' fraction for 11 trace elements (Meyer, 1991).

The definition of contaminated soils has become a crucial question, especially in land deals which legally oblige the vendor to inform the vendee of any element that may indicate a contamination. Use of local criteria, taking into account the soil and geochemical background would be more satisfactory than considering a single formula as in the Netherlands. Mobility and toxicity of the elements depend on their speciation and eventually on many parameters of the pollutant–soil–organisms system. Significant progress with respect to this problem for evaluation of contaminated sites would therefore be attained by considering values other than the total contents. To the threshold values for total concentrations could be added the threshold values for the soluble or bioavailable fractions.

CONCLUSION

The surface horizons of soils form a component of the biosphere in which there is often an accumulation of elements dispersed in small amounts in the earth's crust or concentrated in deposits. These local or diffuse accumulations are most often due to human activity. This creates problems with respect to quality of the environment and food security because many of these elements, beyond a certain threshold of concentration, induce sublethal or acute toxicities in individuals and even a genotoxicity in populations. Solutions appear difficult to implement because these elements are not degradable, the toxicity thresholds are often low and the chemical properties complex.

The most serious current contaminations are local. Their treatment needs to be carefully considered, taking into account the biogeochemical cycles of the elements, soil uses, impact of these 'black spots', whether treated or left as is, and of the treatments themselves on adjacent ecosystems.

To avoid diffuse pollution and to preserve soil quality, durable solutions must be encouraged on the one hand, for the use of soils in agriculture and in industrial and urban zones, and on the other, for management of wastes and dumping: sewage sludges, animal slurries, household wastes, and industrial discharges in water and the atmosphere.

Glossary

Decile	In a series of measurements, threshold values that correspond to the distribution of ordered and regrouped results by tenths (the first decile corresponds to the 10% lowest values, the fifth decile is equal to the median, etc.).
Genotoxicity	A harmful effect manifested in the genome (e.g.: mutation, chromosomal aberration, ...).
Heavy metal	A metal element with a high atomic number, likely to form very slightly soluble sulphides.
Ligand	In chemistry, a simple or compound body likely to form bonds by using part of its external electrons.
Metallophyte	A plant that grows on substrates rich in metals and accumulating them (> 0.10%) in its tissues.
Micronutrients	An element essential in small quantity for survival or development of a living organism.
Phytotoxicity	Toxicity to plants.

Rhizobia   A group of bacteria capable of assimilating nitrogen from the air when they live in symbiosis with certain plants (at the root level).

Trace element   An element normally found in low concentration in a natural environment or in the tissue of an organism.

REFERENCES

Baize D. 1997. Teneurs totales en éléments traces métalliques dans les sols (France). INRA Editions, Paris, 410 pp.

Buffle. 1988. Complexation Reactions in Aquatic Systems: An Analytical Approach. John Wiley & Sons, Chichester, NY.

Cambier P. 1994. Contamination of soils by heavy metals and other trace elements: a chemical perspective. Analysis Mag. 22: 21-24.

Coïc Y, Coppenet M. 1989. Les oligo-éléments en agriculture et an élevage. INRA Editions, Paris.

Giller KE, Nussbaum R, Chaudri AM, MacGrath SP. 1993. *Rhizobium meliloti* is less sensitive to heavy-metal contamination in soil than *R. leguminosarum* bv. *trifolii* or *R. loti*. Soil Biol. Biochem. 25: 273-278.

Godin. 1983. Les sources de pollution des sols. Essais de quantification des risques dus aux éléments-traces. Science du Sol 2: 73-87.

Juste C, Chassin P, Gomez A, Lineres M, Mocquot B, Feix I, Wiart J. 1995. Les micropolluants métalliques dans les boues résiduaires de stations d'épuration urbaines destinées à la fertilisation des sols agricoles. ADEME Angers.

Juste C, Mench M. 1992. Long term application of sewage sludge and its effects on metal uptake by crops. In: Adriano DC (ed.). Biogeochemistry of Trace Metals. Lewis, Boca Raton, FL, pp. 159-193.

Lobinski R, Witte C, Adams FC, Teissedre PI, Cabanis JC, Boutron CF. 1994. Organolead in wine. Nature 370: 64-68.

Logan TJ. 1992. Reclamation of chemically degraded soils. Adv. Soil Sci. 17: 13-35.

Loué A. 1993. Les oligo-éléments en agriculture. SCPA, Aspach-le-Bas, Nathan, Paris.

Meyer K. 1991. La pollution des sols en Suisse. Programme national de recherche Sol. Liebefeld-Berne.

Ministère de l'Environnement. 1994. Recensement dess sites et sols pollués. Direction de la prévention des pollutions et des risques. Service de l'environnement industriel, Paris.

Ross S. 1994. Toxic Metals in Soil-Plant Systems. John Wiley & Sons, Chichester, NY, 469 pp.

Vansgronsveld J, Van Assche F, Clijstèrs H. 1994. Reclamation of a bare industrial area contaminated by non-ferrous metals: *in situ* metal immobilization and revegetation. Environ. Pollut. 87: 51-59.

Yassoglou N, Kosmas C, Asimakopoulos J, Kallianou C. 1987. Heavy metal contamination of roadside soils in the greater Athens area. Environ. Pollut. 47: 293-304.

Part IV

Better Soil Management

CHAPTER 12

Use of Pesticides and Fertilisers in Agriculture

C. Clermont-Dauphin, J.M. Meynard

THE ISSUES

The considerable increase in agricultural productivity recorded in Western Europe over the past 40 years (Fig. 12.1) would not have been possible without extensive use of fertilisers and pesticides as well as improvements in varietal performance. With the help of these tools Europe first became self-sufficient, then an exporter, and now finds itself confronted with the problem of overproduction. In many tropical countries, on the other hand, where the level of inputs remains low, agricultural output is severely limited by low levels of assimilable minerals in the soil and a high incidence of pests, weeds and diseases.

Nevertheless, despite their undeniable role in increased production, the use of fertilisers and pesticides in agriculture now clearly needs to be re-examined, not only because of the fragile economics of small farms but also because of the environmental degradation now occurring in intensive farming regions, both in Western Europe and in tropical countries, where the use of chemical inputs is now more widespread. The damage includes water polluted with nitrates and pesticides,

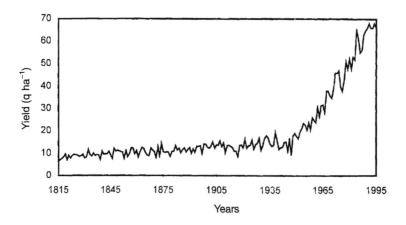

Fig 12.1   Increase in wheat yield in France

reduced biodiversity and, especially in tropical regions, the declining fertility of some soils.

Nonetheless, chemical inputs are still necessary. At the global level, the growth of the human population requires a parallel growth in agricultural output. The chronic problem of undernourishment in some 'underdeveloped' countries can only be solved by increasing agricultural output in those countries. Simply transferring surpluses from North to South is not a solution for the medium or long term, given its consequences for southern small landholders. In Europe, falling prices for many farm products as a result of changing agricultural policy will probably lead to extensification. But ensuring a regular supply of food at low prices is not compatible with the total elimination of chemical inputs. The long periods (about 1 year) of set-aside introduced by the European Agricultural Policy as a means of reducing productive area, rather than restoring fertility as the image of fallowing would suggest, actually tends to favour the spread of parasites and weeds, so that pesticide use has to be increased on the subsequent crop (Sebillotte et al., 1993).

This chapter examines environmental problems linked to pesticide and fertiliser use in different types of agriculture and proposes ways to limit the damaging effects of misuse of these inputs on soil fertility and the environment.

STRATEGIES IN HIGH-INPUT FARMING: EXAMPLE OF NORTH-WEST EUROPE

Limiting Risk of Groundwater Pollution by Nitrates

How does nitrate pollution occur?

The nitrate ion is the dominant mineral form of nitrogen in cultivated soils of temperate regions. It is also the form of nitrogen most readily taken up by most plants, which need large amounts of nitrogen to build leaf area and reproductive parts and thereby their yields. This justifies the use of mineral and organic nitrogenous fertilisers to supplement the soil nitrogen supply and prevent major deficiencies. Groundwater pollution by nitrate occurs when two factors combine:

- The presence of residual nitrate in the soil, which has not been absorbed by the crop. This nitrate may come from earlier fertiliser application or mineralisation of organic matter.
- A water balance in surplus, i.e., water has been supplied by precipitation and/or irrigation in excess of total evaporation from the soil and transpiration from the crop.

It is therefore wrong to attribute nitrate pollution exclusively to the application of mineral fertiliser: if fertiliser is applied in a rational manner—according to the ability of the crop to absorb nitrogen, and avoiding periods when the soil is draining—there is no risk of pollution. On the other hand, application of organic manures may cause pollution: large quantities of manure on irrigated vegetables and urban compost in vineyards are more harmful to the environment than a balanced input of mineral fertiliser. In intensive animal husbandry, a grassland used for grazing is far more prone to nitrate pollution than one which is mown, because of the extremely high concentration of nitrogen in places where animals gather; the amounts of dung and urine in these areas far exceed the absorption capacity of the vegetation.

Conversely, zero use of mineral and organic fertilisers does not necessarily eliminate the risk of nitrate pollution. For example, a part of the nitrogen fixed by symbiotic rhizobia remains in the soil after the legume is harvested and may be leached into the groundwater.

Thus there is no simple solution to the problem of groundwater pollution by nitrates: results will only be achieved by detailed study of the factors that may help to identify *'risk situations'* and selecting suitable methods to deal with each.

Identifying risk situations

The **nitrogen balance** is an important criterion for recognising pollution risk situations. If inputs of nitrogen to the soil (mineral and organic fertilisers, fixation of atmospheric nitrogen by symbiosis between rhizobia and legumes etc.) are greater than the outputs (nitrogen contained in harvested crop or in grass eaten by animals etc.), the excess may enrich the soil in the short term, but the nitrogen-holding capacity of a soil is limited and mainly depends on its humus content. In the medium term, part of any excess in the nitrogen balance will find its way into drainage water, especially in shallow, permeable soils such as the shallow calcareous soils (rendzinas) found in the Aunis region and the Champagne Berrichone in France. In soils of this kind, most of the nitrate ions present in the soil in the autumn are carried down to a depth of more than 50 cm over the course of a single winter, by water percolating through the soil. And at 50 cm depth, they are often beyond the reach of crop roots. In a deep loamy soil, on the other hand, the roots of annual crops may reach a depth of about 1 to 1.2 m and the nitrogen is not leached beyond this depth in a single winter (Fig. 12.2). In these soils, if the previous crop has left nitrate in the ground due to root disease or dryness, the residual nitrate can be deducted from the fertiliser input rate for the subsequent crop; this will prevent pollution without reducing crop production.

Currently available models of solute transfer can be used to estimate the probability, in a given type of soil, that nitrates in the topsoil (produced by biological activity or introduced by fertilisers) will be leached beyond rooting depth. This probability directly depends on the amount of percolating water. Rain and irrigation are therefore also factors to be taken into account for assessing risk situations. Thus in irrigated horticultural systems, such as are often found near rivers on sandy alluvial soils, excess fertiliser very soon reaches the water table.

Leaving soil bare during the leaching period increases the risk of nitrate pollution: crops absorb nitrate from the soil and prevent it from leaching, while transpiration reduces the amount of percolating water (Table 12.1). With arable crops, the most high-risk rotations are those that leave the soil bare in winter: maize as a monoculture, for example.

What farming practices will help control nitrate pollution?

The central point in a strategy to limit pollution is therefore to **carefully adjust the amount of soil nitrate from mineral and organic fertiliser to the crop's requirements**. This means taking great care to choose fertiliser rates and application dates according to the developmental

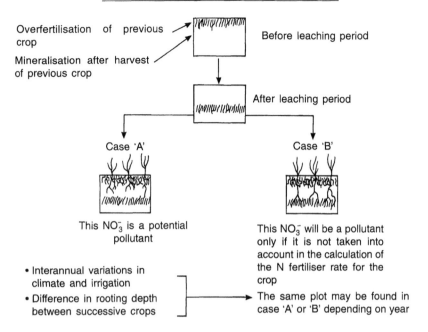

Examples: 2 extreme situations

	Examples (conditions in France)	Determinants of pollution risk
'A' highly predominant	• Shallow soil, arable crops • Irrigated vegetable crops regardless of soil type	Quantities of NO_3^- present in the soil every year during the leaching period
'B' highly predominant	Deep soil, non-irrigated arable crops	Nitrogen for next crop

(after J.M. Meynard)

Fig. 12 2 Nitrate leaching risk depending on environment and cropping systems.

Table 12.1 *Amounts of drained water and leached nitrogen in two types of soil, with wheat-maize rotations; comparison between sowing a catch crop (mustard) and leaving a bare soil (Chapot, 1990)*

Loam				Silty sand			
Catch crop (mustard)		Bare soil		Catch crop (mustard)		Bare soil	
Water (mm)	N (kg ha^{-1})	Water (mm)	N (kg ha^{-1})	Water (mm)	N (kg ha^{-1})	Water (mm)	N (kg ha^{-1})
56	1	144	60	71	10	163	118

stage of the crop (in wheat, for example, the beginning of stem elongation is considered an important moment for applying nitrogenous fertiliser), and weather patterns (avoiding periods when the water balance is in surplus and periods when the soil is dry, as the fertiliser will then not be diffused toward the roots). With today's improved knowledge of soils and agricultural potentialities, fertiliser application can be adjusted case by case, plot by plot (Precision Agriculture). In this way the amount of fertiliser used in wheat crops in the same region may vary from less than 50 kg N per hectare on a soil with a high nitrogen content from earlier crops (for example peas, alfafa or grassland) or from recent organic supplies, to more than 250 kg N per hectare on a plot where a high yield is expected and the previous crop's nitrogen uptake was efficient, leaving very little available nitrogen in the soil.

However, although quite high-quality fertilisation models are now available for most crops in Western Europe, it is not always possible to predict the conditions for a crop's fertiliser uptake with complete accuracy. Pests, diseases or a prolonged drought, among other factors, may affect nitrogen absorption by plants and leave behind large amounts of nitrate in the soil after harvest. Agronomists are currently working to help farmers identify these situations and have recommended that they avoid leaving the soil bare in winter by sowing a catch crop, not for harvesting but to trap the nitrate and prevent its leaching.

Other cropping techniques also impact on nitrate pollution: irrigation must be used sparingly to limit the risk of water balance surpluses; tillage, rotations and management of crop residues also influence nitrogen cycling and have an indirect effect on nitrate leaching.

Control of Pests and Diseases with Minimal Pesticide Treatments

Reasoning behind systems that use pesticides intensively

Losses due to insects, pathogens and weeds are estimated at 35% of agricultural production of the world. If we add post-harvest losses due to various micro-organisms, insects and rodents, the total loss can reach 45% (Vincent and Coderre, 1992).

The chemical method of crop protection has been increasingly employed since the 1950s because of its relative simplicity and apparent effectiveness. However, the medium-term efficacy of chemical protection has been questioned, in view of the development of populations of pathogens, weeds or pests resistant to the most frequently used active ingredients, and the unintentional destruction of other components of the agrosystem, including the natural enemies of crop pests.

Changes in wheat management in France over the last twenty years clearly show how important a role pesticides play in present intensive cropping systems. Although pesticides have been used on wheat for decades, there was a sharp increase from the beginning of the 1980s. Until that time, the advice farmers received was to eliminate the major limiting factors. Attention focused successively on controlling lodging (with resistant varieties and growth regulators), diseases (with fungicides), nitrogen nutrition (the N balance sheet method was proposed and tested during the 1970s) and pests (with insecticides against aphids during grain filling). Various herbicides theoretically enabled farmers to control the main weeds. Once these limiting factors were under control, the final limiting factor for yield was the photosynthetic activity of the crop.

The movement towards intensification in the early 1980s was based on the desire to combine available methods of control over limiting factors with methods designed to achieve the highest possible photosynthetic potential in the crop. Farmers were advised to advance sowing dates so as to increase the duration of photosynthesis, and to increase planting density to intercept more photosynthetically active radiation at the beginning of the cycle. In the 10 years from 1975 to 1985, in the Paris region, the frequency of sowing before the 20th of October increased from about 30% to almost 60%. For the same planting date, the quantity of seed used per hectare increased by more than 30% (Meynard, 1991).

Early planting, however, increases the risks of autumn emergence of graminaceous weeds, diseases such as take-all, eye-spot or septoria, lodging and attacks of aphids, which are vectors of viruses. Dense sowing increases risks of lodging and disease (eye-spot, take-all...). So these technical changes were not possible before effective plant protection chemicals became available. To reduce the risks entailed by early sowing, farmers had to undertake systematic weeding in autumn, increase fungicidal protection and often spray insecticide in autumn as well as in spring. This cropping system, commonly known as *'intensive wheat cultivation'*, is heavily dependent on the use of the pesticides that made it possible. It is the main factor responsible for the increase in pesticide levels generally recorded in surface and groundwater in arable regions of France.

Development of integrated pest management (IPM)

Obviously, a substantial reduction in the use of pesticides cannot be achieved without changes in the entire cropping system. For example, with wheat crops sown early and densely with large amounts of fertiliser, the disease risk is such that a sharp reduction in fungicide application may lead to major production losses.

To reduced pesticide use, measures should be taken to control the entire life cycle of the parasites: this is the principle of *integrated pest management*. It may involve several measures:

- reducing the conservation, spread or multiplication of the parasite inoculum,
- encouraging the crop pests' parasites or natural predators,
- using varieties resistant to or tolerant to parasitical damage.

In the case of wheat, today's lower prices encourage the use of less intensive systems. Fungicide treatments can be considerably reduced, without increasing variability in yield and gross margins, by choosing varieties resistant to cryptogamic diseases and effecting a co-ordinated reduction in sowing densities and nitrogen fertiliser rates (Meynard and Girardin, 1992). High yields can also be achieved with little fungicide input by associating in the same stand different varieties with complementary disease resistances. Applications of insecticides and fungicides may be based on models that forecast epidemics and so make it possible to intervene at the most opportune moments. The insecticides chosen should be harmless to auxiliaries (for example, entomophagous insects, such as the Coccinellidae, which limit proliferation of aphid populations).

In these ways the combined efforts of researchers and farmers should be able to reduce modern agriculture's dependence on pesticides. The implementation of integrated pest management worldwide will be the only way to obtain the high yields necessary to feed growing populations while minimising environmental damage (accumulation of pesticides in the soil, pollution of groundwater and rivers, reduced biodiversity etc.).

STRATEGIES IN LOW-INPUT AGRICULTURE: EXAMPLE OF HUMID INTERTROPICAL ZONES

In a humid tropical environment, the weathering of minerals in the soil together with heavy leaching of the weathered products, result in a predominance of ferrallitic soils with very low levels of minerals that crops can assimilate. These soils have a low cation exchange capacity, i.e., a very weak capacity to retain cations such as K^+ and Mg^{++}; the cations are therefore leached out in drainage water. Phosphorus fixed by hydroxides of iron and aluminium remains in the soil, but in a form that cannot be absorbed by the plants.

Traditional Fertility Management Practices

The management of soil mineral fertility is therefore an important factor for viable, low-input tropical agriculture. In most cases the rule is farm-level self-sufficiency, different cultural practices ensuring a turnover of minerals and a minimum provision for each year's crops. In Haiti, for example (latitude 20°N, longitude 75°W), nutrients are generally integrated into tilled horizons of the soil in three ways (Fig. 12.3) (Bellande et al., 1980; Pillot et al., 1994):

- By non-tilled fallows, often left for grazing for one to three years. During this period, volunteers of crop species and various weeds absorb the minerals from mineralisation of organic matter or from rainfall; several wild legumes fix atmospheric nitrogen. The organic matter produced in this way and ploughed in before planting the crop will supplement the crop as it becomes mineralised.
- By transferring crop residues or manure from distant plots to those near home, where shorter rotations are practised. In this way nearby plots are enriched at the expense of the more distant.
- By combining perennial and annual species. Such combinations are characteristic of the plots close to the family home. The trees of this 'garden' transfer fertility from deeper soil layers to surface

Fig. 12.3   Traditional fertility management practices in low-input agriculture. The case of Haiti. 1—Rise of minerals from deeper soil horizons. ... 2—Transfer of crop residues.

layers. This plot also receives kitchen wastes and some of the residues from distant plots.

Similar practices are found in most farming systems with limited inputs (Houdard, 1993). In forest regions, slash-and-burn (planting after burning the forest) is based on the same logic: setting fire to a forest every 10 to 20 years enriches the soil with easily assimilable minerals, increases pH and destroys or inhibits weed seeds near the surface. This initial fertility is exhausted after two or three years of cropping; the plot is then abandoned until the secondary forest has fully regenerated.

However, where population pressure is high, as it now is in most tropical countries, such practices are becoming increasingly unsuitable for two reasons: (i) they take up space; (ii) they mainly involve transferring part of the minerals extracted by crops from one plot to another. Thus they cannot significantly improve productivity per unit of arable area.

Effects of Mineral Fertilisers on Soil Fertility

An external supply of minerals is therefore necessary. Recently, the ready availability of chemical fertilisers has enabled farmers to increase their use. However, because ferralitic soils have a low storage capacity for assimilable minerals, mineral fertilisers will not greatly enrich the soil. The fertiliser will have a mainly short-term positive effect on the crop it is applied to, and on the following crop. A more long-term effect can be obtained by supplying organic matter. As a rule, however, little organic matter is available on small tropical farms.

Sometimes, the use of mineral fertilisers may cause a damaging loss of fertility in ferrallitic soils in the long term. This may result in two situations:

Acidification of the soil

This is caused by: (i) increased leaching of cations associated with the supply of 'accompanying anions' NO_3^-, SO_4^-, Cl^-; (ii) release of H^+ ions by nitrification of ammonia fertilisers; (iii) adsorption of anions accompanying chemical fertilisers on the sesquioxides of iron and aluminium. An example of acidification of ferralitic soils in Burkina Faso, associated with use of fertilisers containing nitrogen, phosphorus and potassium in long-duration field trials, is given in Table 12.2. Acidification was accompanied by an increase in exchangeable aluminium levels leading to aluminium toxicity and lower sorghum yields.

Table 12.2   *Evolution of function of NPK fertilisation, yield of sorghum, of pH and content of exchangeable Al^{3+} of a soil after 10 years (1969/1979) of sorghum monoculture (Piéri, 1989)*

Period 1969-79	Control (without fertiliser)	Moderate fertilisation	Heavy fertilisation
Reduction in yield (kg ha^{-1})	45	215	579
Fall in pH	0.1	0.5	0.3
Increase in exchangeable Al^{3+} (meq 100 g^{-1})	0.01	0.43	0.41

Reduction of minor cations in the soils due to their displacement by cations supplied by fertilisers

Displacement of magnesium by potassium is a probable cause of the negative effect of mineral fertilisation on bean yields, as seen in Haiti on ferrallitic soils. The negative responses to fertilisers are accompanied by a statistically significant decrease ($P = 2\%$) in the absorption of magnesium by the bean plant, while absorption of potassium is not affected (Table 12.3). These negative responses of the bean yield are mainly found in situations where the previous crop received a large amount of potassium fertiliser (Clermont-Dauphin, 1995).

Table 12.3   *Yield, and absorption of K and Mg in plots where bean response to fertiliser was negative (average of 17 trials) (Clermont-Dauphin, 1995)*

Treatment	Without fertiliser	With fertiliser
Average yield (kg ha^{-1})	1071	915
Absorbed K (g m^{-2})	4.14	4.75
Absorbed Mg (g m^{-2})	0.50	0.42

Changes in Other Cropping Techniques Related to Application of Mineral Fertilisers

Should the application of mineral fertilisers always go hand in hand with the use of pesticides and new varieties suitable for intensification, as in the technical package of the Green Revolution? This is by no means certain. Traditional cropping systems, though not very productive, are generally well suited to local physical conditions (soil, climate) and farmers' financial resources. In Haiti, for example, to maximise fertiliser use efficiency it was decided to test the usefulness of applying fertilisers in experiments with a genotype selected for its productivity and resistance to several leaf diseases prevalent in the region. It appears, however, that the farmers' plots where these trials were conducted were managed by short rotation in which the bean crop recurred frequently; the new variety was heavily infested by a root disease that curtailed absorption of fertiliser by the crop. By contrast, an association of varieties already used by farmers, grown under the same conditions, suffered less infestation because some varieties of the association were resistant to the disease. The farmers' technique of using a mixture of varieties therefore provided a higher level of yield stability in the face of parasite risk and a better guarantee of fertiliser response than a pure stand of a highly productive but non-resistant variety.

More generally, the genetic homogeneity that results from the use of selected varieties is more favourable to the development of epidemics; the genetic diversity in mixtures of varieties or species, often observed in low-input agriculture, may be an effective way to limit the damage caused by many parasites.

CONCLUSION

As shown by several of the above examples, the soil is a 'memory' of the effects of cropping systems: residual mineral nitrogen, residues of pesticides, seed banks of weeds, parasitic inocula, retention and release of cations etc. Attempts to obtain immediate profitability sometimes—too often, indeed—lead to overlooking these factors, underestimating their positive consequences and forgetting the negative until irreversible changes have occurred. To improve the sustainability of cropping systems, a better appreciation of this soil 'memory' is required.

In view of the diversity of soils, climates and socioeconomic conditions, solutions should be designed on a case-by-case basis. The days of standardised systems are over. The Green Revolution in subtropical regions and the intensification of arable crops in the 1980s in Western Europe had a negative impact on the environment and sometimes on the income of those who adopted them, because they were recommended and adopted without reference to the diversity of soils, farms and climates, and without paying attention to their environmental impacts.

Agricultural production depends on ecosystems that function in far more complex ways than some agronomists would like to believe. Chemical inputs should be regarded as one of the means, but not the only one, for managing farm ecosystems.

Glossary

Agricultural potentialities	Maximum productivity of a plot taking into account its soil, microclimate and available techniques.
Cation exchange capacity (CEC)	Number of negative charges likely to retain cations (for example: NH_4^+, Ca^{2+}, K^+, Mg^{2+}, H^+, Al^{3+}...) per unit of soil weight.
Cropping system	All agricultural practices adopted on a plot (crop rotations, sowing, tillage, fertilisation...).
Disease inoculum	Generic term characterising any element of a parasite capable of contaminating a host .
Fallow	State of a plot between harvesting one crop and sowing another.
Ferrallitic soil	Typical soil of humid tropical zones, considerably weathered by climatic conditions, rich in iron and aluminium oxides and with a low CEC.
Green Revolution	An agricultural policy which in the 1970s and 1980s promoted intensification of plant production in tropical areas, based on selected varieties, fertilisers and pesticides. The Green Revolution helped to increase food production in many countries.
Inputs	Materials spread on a plot to improve its productivity (mineral and organic fertilisers, pesticides, growth regulators etc.).
Rhizobium-legume symbiosis	Association of a bacterium (rhizobium) and a plant; the bacteria fix atmospheric nitrogen, of which a part is used by the legume; due to this symbiosis, legume crops do not require nitrogenous fertiliser.
Sustainable agriculture	Agriculture that responds to present requirements without compromising the ability of future generations to respond to their own requirements (Conference, Rio de Janeiro, 1992).

REFERENCES

Bellande A, Brochet M, Cavalié J, Fourcault H, Mondé C, Pillot D, de Reynal V 1980. Espace rural et Société agraire en transformation. Recherches haïtiennes, n°2, Haïti, 172 pp.

Chapot JY. 1990. Azote, la juste dose. Cultivar 279: 35-36.

Clermont-Dauphin C. 1995. Étude de la fertilisation minérale d'une association haricot-maïs en zone tropicale humide. Diagnostic des effets des systèmes de culture. Thèse doc. Eng. INA P-G, Paris.

FAO. 1994. Agriculture: Horizon 2010. Collection FAO.

Fraval A. 1993. La lutte biologique. Les Dossiers de la Cellule Environnement de l'INRA, N°5, 238 pp.

Houdard Y. 1993. Agricultures dans les hautes collines du Népal central. INRA, Département de Recherche sur les Systèmes Agraires et le Développement, 291 pp.

Meynard JM. 1991. Pesticides et itinéraires techniques. In: Protection des Plantes, Phytosanitaires et Biopesticides. P. Bye, C. Descoins, A. Deshayes (coord.) INRA Ed., Paris, pp. 85-100.

Meynard JM, Girardin P. 1992. Produire autrement. Courrier de la Cellule Environnement de l'INRA, n°15, 19 pp.

Piéri C., 1989. Fertilité des terres de savane. Min. Coop. CIRAD-IRAT, 443 pp.

Pillot D et al. 1994. Paysans, Systèmes et Crise. Travaux sur l'agraire haïtien. SACAD-FAMV, T. 3.

Sebillotte M, Allain S, Doré T, Meynard JM. 1993. La jachère et ses fonctions agronomiques économiques et environnementales. Diagnostic actuel. *C.R. Acad. Agric. Fr.* 4: 105-121.

Vincent C, Coderre D. 1992. *La lutte biologique*. G. Morin (ed.), 661 pp.

CHAPTER 13

Purifying Functions of soil and their Limits

J.C. Germon

The quality of water that migrates to aquifers depends on the capacities of the soil and subsoil through which it passes to eliminate the pollutants. The mechanisms that enable transformation of surface water with a fairly high content of suspended matter and dissolved organic or mineral elements into potable water constitute the purifying properties of the soil.

Man has long used this purifying effect of the soil to regenerate the quality of waters he has dirtied: in ancient Athens waste waters were spread on grasslands. During the nineteenth century improvements in hygiene led to the establishment of fields for spreading waste waters from major agglomerations such as Paris, London or Berlin. Those in Achères, west of Paris, were still in operation until recently. In the twentieth century, this type of treatment was supplanted in developed countries by industrialised procedures that require less space.

In the USA, since World War II and in Europe, since the 1970s, interest in the use of soils or rustic systems derived from soil for treating water or different wastes has revived. Treatment by soil may prove to be a more effective method for waste waters from small villages than by more sophisticated methods which are ineffective at this level. In other cases, it can be an inexpensive means of final treatment, enabling recycling to the aquifers of water treated earlier at a treatment station. In the case of some effluents, especially those from agrofood industries, it may also be a means of utilising the fertiliser elements contained in the spread waters and wastes.

Nevertheless, the purifying capacity of the soil is limited by its own functional characteristics:

- the exchange properties of the absorption complex will determine the quantities of mineral elements that can be retained temporarily and later returned and made available for vegetation;
- the biological characteristics of soil determine the capacity for biotransformation and elimination of the organic matter supplied;
- ambient physical conditions, in particular temperature, will determine the survival capacity of introduced pathogenic microorganisms.

These treatment capacities are also limited by thresholds in the concentration of various harmful or toxic elements that may accumulate in soil and plants and are likely to move in solution to potable water.

Each of these purifying functions is examined here and their limits specified.

FILTRATION AND SOIL CLOGGING

Soil is a porous medium which allows a fairly easy circulation of air and water and retains suspended particles on its surface. The quantity of water treated by spreading over a soil is very often regulated by its infiltration capacity. This varies on the one hand with structure of the soil and its initial porosity and, on the other, with the extent of clogging subsequent to spreading.

Soil porosity depends on its texture, especially its clay content, and on its more or less compact and stable structure. Infiltration capacity can be characterised in the first instance by measurement of permeability expressed by Darcy's law. It shows that a soil of medium texture may allow infiltration to saturation point of several centimetres of water per hour. But this infiltration rate varies widely (Table 13.1). When the rate is very high, the period of contact between the dissolved elements and the active surface of the soil may be too brief to ensure an adequate purification. When the infiltration rate is very low, the soil will soon be saturated with water, leading to the phenomenon of surface runoff: the polluted water will then directly rejoin the surface outlets, ditches and

Table 13.1   *Variation in infiltration rate of a soil expressed in terms of Darcy's law*
(K = Q/Si; where Q is the quantity of water transported across a column of
soil of section S submitted to a hydraulic gradient i) (Chamayou and Legros,
1989)

Infiltration rate	K in cm s^{-t}	K in cm h^{-1}
Very slow	$< 3 \times 10^5$	< 0.1
Slow	3×10^5 to 1.5×10^4	0.1 to 0.5
Fairly slow	1.5×10^4 to 6×10^4	0.5 to 2.0
Moderate	6×10^4 to 1.7×10^3	2.0 to 6.5
Fairly rapid	1.7×10^3 to 3.5×10^3	6.5 to 12.5
Rapid	3.5×10^3 to 7×10^3	12.5 to 25
Very rapid	$> 7 \times 10^3$	> 25

later streams, without having been treated. This leads to selection of soils to be used for such purposes depending on the quantity of water to be treated. Clay or silty soils may be suitable for small volumes with a high content of pollutants. Soils with a moderate texture are most suitable for quantities corresponding to the amounts normally used for irrigation in agriculture. Sandy soils are preferable for spreading waters with a low content of pollutants and especially for filtration of urban waste waters that may have already been subjected to earlier treatment and for which amounts of 500 mm day^{-1} are usually provided.

When spreading waste waters on soil a more or less rapid clogging generally occurs, evidenced by the permanent presence of water on the surface, sudden reduction in the rate of infiltration, lower level of performance with respect to purification and the appearance of anaerobic phenomena. If the supply is temporarily stopped and the soil left to rest to enable the surface layers to dry up and air to penetrate the pores again, permeability may be restored as indicated in Fig. 13.1.

The clogged layer is generally superficial and does not exceed a few centimetres. It may be due to an accumulation of organic matter from the effluents or result from development of a proliferating microbial biomass. This clogged layer may also be characterised by an accumulation of various mineral elements from the effluents or be due to the presence of sulphides formed in the absence of oxygen and imparting a dark colour that disappears when the soil is aerated again.

Clogging may also be caused by alteration of soil structure due to the composition of the water supplied, notably its high content of sodium salts, as observed in some industrial waste waters which are able to disperse clays.

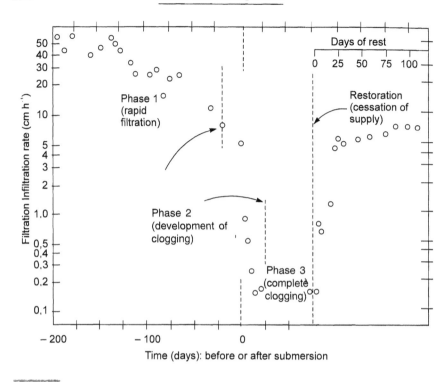

Changes in permeability of a sandy soil subjected to spread of urban waste waters: indication of soil clogging and restoration of permeability (after Thomas et al., 1966).

Clogging which may temporarily improve the filtration effect of a soil due to better retention of suspended particles, eventually alters the purifying properties. It is therefore desirable to prevent soil clogging and to correct it when it occurs. For this reason effluents with an excessively high salt content, in particular of sodium, are to be avoided. When clogging is organic, the soil can be worked on the surface. Thus in filtration basins with sandy soils used for treating urban waste waters, raking and removal of the organic film forming a coating on the surface, which is easy to destroy upon drying, enables restoration of permeability.

SOIL: A PHYSICOCHEMICAL REACTOR

The composition of water which passes through a soil changes according to the physicochemical transformations of the elements contained during contact with the organomineral complex: reactions of precipitation-solubilisation and exchanges mainly with the absorption complex.

This leads to a more or less high degree of mobility of the elements supplied.

The main elements that regulate these physicochemical transformations are:

- The exchange capacity of the absorption complex which helps to retain the ionised elements from the solution on the organomineral matrix of the soil by exchanging them with the ions initially present on the absorption complex. For cations, the exchange capacity varies from a few milliequivalents[1] per 100 g sandy soil to 60 to 80 meq per 100 g clayey soil, with average values of 15 to 25 meq in most French soils. It selectively retains cations with an absorption efficiency that increases in the order of their valence (monovalents < bivalents < trivalents). For a particular valence, the increasing order is Li < Na < K < Rb < Cs for monovalents, the retention of ammonium being similar to that of potassium, and Mg < Ca < Sr < Ba for bivalents. The exchange capacity of anions, generally lower than that of cations, is only a few meq 100 g^{-1} and preferentially retains polyvalent anions (PO_4^{3-}, SO_4^{2-}), while monovalent anions (NO_3^-, Cl^-) are almost not retained.

- The pH: this regulates the solubility of most mineral elements. In soil, many elements are found as hydroxides or carbonates; their precipitation and consequently non-solubilisation are largely dependent on an alkaline pH, while an acidic pH assists in their solubilisation.

- Redox phenomena: the extent of soil aeration determines its oxidoreduction state which regulates the balances between oxidised and reduced compounds of certain elements, especially between oxidised and slightly soluble forms of iron and manganese (Fe^{3+} and Mn^{4+}) and their reduced and much more soluble forms (Fe^{2+} and Mn^{2+}). The change from a poorly aerated and reducing environment to a well-aerated and oxidising one is accompanied by the process of precipitation of these elements which may concomitantly trap and retain other compounds. Moreover, in a highly reducing environment, reduction of sulphur compounds may be observed together with the formation of practically insoluble sulphides, which cause precipitation of the elements bound with them.

On the whole, discharge of waste waters or polluting products on the surface of a soil results in a qualitative and quantitative change in

---

[1] The exchange capacity of soil is usually expressed in milliequivalents, or meq per 100 g dry matter.

composition of the water percolating through the soil. The most soluble elements, such as chlorides and nitrates, move with the water while those likely to be absorbed or precipitated, such as phosphates, calcium or potassium, remain localised in the surface horizons. Elements that are poorly retained, such as sodium, will behave in an intermediate manner and descend slowly down the profile. The same applies to organic molecules which migrate fairly easily depending on their ionisation level. Thus some organic micropollutants cannot be biodegraded in soil because they migrate too rapidly. Adsorption of these micropollutants on the constituents of the soil, by ensuring their retention, is a necessary prerequisite for their biodegradation, provided they can be desorbed or are biodegradable in the adsorbed form.

Duration of retention by soil varies according to the elements and mechanisms involved. While adsorbed mineral products may be readily utilised by microflora or vegetation, precipitation phenomena cause a more durable immobilisation. Adsorption of organic products may change over time and lead to the formation of bound and non-extractable products. This is well known for some organic micropollutants such as triazines.

Retention of mineral products by the soil may be accompanied by accumulation of both fertiliser elements (Fig. 13.2) and undesirable elements. Repeated spreads of liquid manure are accompanied by increased concentrations of copper and zinc. Thus it has been estimated that the threshold of toxicity for accumulated copper may be reached in the soils of Brittany in little more than a century if the quantities used and the composition of these liquid manures remain the same. These phenomena of accumulation are being carefully monitored for the principal heavy metals (Cu, Zn, Pb, Cd, Hg, Ni, Cr) and for selenium, which is likely to be introduced by the spread of urban sewage sludges. Concentration norms have been established for these elements and must be complied with for both the sludges and the soil used (Table 13.2).

SOIL: A BIOTRANSFORMER

When waste waters or residues are spread over soil, the intention is to ensure biodegradation of the organic matter by soil microflora. This organic matter will serve as a source of energy and elements that constitute the microbial biomass. All these transformations lead to (1) complete oxidation of part of the compounds to mineral products (CO_2, H_2O, NO_3^-, SO_4^{2-}, PO_4^{3-}); (2) formation of microbial cell constituents; (3) release of incompletely oxidised metabolites (organic acids and other products of fermentation) as well as synthetic products; and

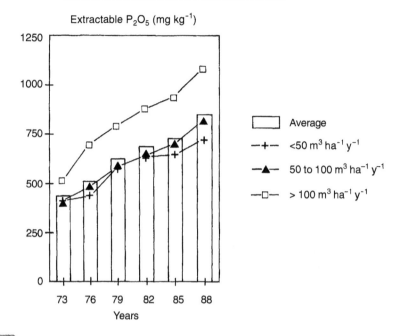

Fig. 13 2   Changes in phosphate content of soils in intensive animal husbandry area with a predominance of piggeries in Brittany, France (1973-1988); effect of amount of slurry spread ($m^3 ha^{-1} y^{-1}$).

Table 13 2   *Maximum values of concentrations of heavy metals for residual sludges (in mg kg^{-1} dry matter) and for soils (in mg kg^{-1} soil) used for spreads. French regulations prohibit the spread of sludges if the content of any single element is higher than the corresponding maximum value, and the use of soil for purposes of spreading if the content of any element is higher than the stipulated maximum value (ministerial decree of 8 January 1998).*

	Maximum values in sludges	*Maximum values in soils*
Cadmium	20	2
Chromium	1000	150
Copper	1000	100
Mercury	10	1
Nickel	200	50
Lead	800	100
Zinc	3000	300
Chromium + Copper + Nickel + Zinc	4000	—

(4) non-transformed residual organic material that constitutes the inherited humic matter. The formed mineral products except for nitrates and organic matter attached to the soil in the form of microbial cells or adsorbed on the clay-humus complex, are considered to be sanitised, while organic matter remaining in solution is considered a potential source of pollution. Therefore the objective in management of biotransformations is to promote reactions that lead to the first series of compounds rather than the second.

Sanitisation of organic products in soil depends on the capacity of the microbial biomass to degrade them and to grow at their expense. This biomass in soil, representing 1 to 4 tons of organic matter per hectare, is contained mainly in the surface horizon, the arable layer, and decreases very rapidly at lower levels as does the rest of the organic matter (Fig. 13.3). This explains why most of the biological purification occurs in the surface soil layer. The activity of this biomass depends firstly on the availability of organic substrates. Without an external energy source, micro-organisms present a slow metabolism, using their reserves or poorly biodegradable products of the humus stock in the soil. The metabolism of micro-organisms meets their maintenance

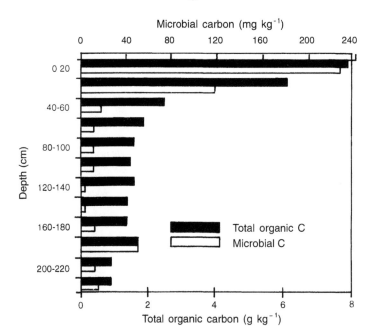

Fig 13 3   Variations in total organic carbon concentration and carbon of the microbial biomass in the profile of a silty soil (source: G. Soulas, INRA, Dijon).

requirements. They do not grow and, on the whole, their population decreases slowly. When a supply of biodegradable organic products is available, the microflora capable of transforming these products become active and grow until a limiting factor appears: this is observed by an increase in the microbial biomass and its respiratory activity. Hence soil microbiologists make a distinction between microflora responsible for slow mineralisation of the humus stock and zymogenic microflora that develops with the supply of an external substrate and may be adapted to this substrate.

The purifying biomass in soil has a considerable ability to increase: at 20°C and in favourable conditions of nutrition, the doubling time for the soil micro-organisms varies depending on the species, from several tens of minutes to several hours or even several days, while the quantities of biotransformed products are related with the synthesised microbial biomass.

The purifying microflora is also highly diversified and adaptable to the products used and conditions of the environment. The high microbial population of soil (10^7 to 10^8 and 10^4 to 10^5 units forming colony per gram of soil in the surface layer from bacterial and fungal microflora respectively) allows detection in a particular soil of micro-organisms capable of degrading a large variety of organic compounds. Compounds such as some simple sugars or organic acids with short chains are completely degraded by a large part of soil micro-organisms while others with less common chemical structures are degraded by highly specialised microflora (Table 13.3). This diversity of micro-organisms and their high reproduction rate are also responsible for their considerable adaptability enabling formation of metabolic capacities for new molecules. As a result, repeated treatments of soils with certain xenobiotics may lead to the formation of a microflora adapted to these products and their quicker biodegradation (Fig. 13.4).

Regulation of the functioning of purifying biomass also depends on factors other than the microflora:

- Biodegradation of molecules or their resistance to biodegradation is first of all due to their chemical structures which allow accessibility to microbial enzymes. This indicates that an organic product spread on the soil will not be purified unless the microflora comprises micro-organisms capable of degrading it; otherwise it may accumulate. The use of non-biodegradable products, or only partly so, such as some chlorinated organic products like lindane has thus been responsible for persistent pollutions.

Table 13 3 *Population levels of various microfloras in the arable horizon of a soil in Dijon, France (source: J.C. Fournier, INRA, Dijon)*

Total microflora	10^8	micro-organisms g^{-1} soil
Microflora degrading 2,4-D	10^2	micro-organisms g^{-1} soil
Microflora degrading MCPA	10	micro-organisms g^{-1} soil
Microflora degrading carbofuran	0	micro-organisms g^{-1} soil

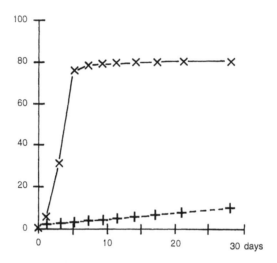

Fig 13 4 Degradation of ^{14}C-carbofuran (3 mg kg^{-1}) in samples of a soil that had never been treated (+: non-pretreated soil) or which had previously received several treatments with this product (x: pretreated soil). Results in % of total quantity of ^{14}C initially introduced. (M.P. Charney, thesis, Univ. Lyon, 1993).

- Development of the purifying microflora in soil also depends on the ability to find all the necessary elements in the environment for cellular synthesis and in particular sources of nitrogen and phosphorus in addition to a carbon substrate. Thus a product spread on soil with a high content of easily transformable carbon compounds may be subjected to slow biodegradation because of a nitrogen deficiency.
- Oxygen is the element that frequently limits the purifying capacity of soil subjected to large amounts of organic matter. In an aerobic environment, microflora growth is much more efficient than in an anaerobic environment and results in complete oxidation of organic matter to CO_2, while nitrogen is transformed to nitrates. In

anaerobic conditions, fermentations lead to an accumulation of incompletely oxidised organic compounds, formation of ammonium from transformation of nitrogenous products and subsequent accumulation of other reducing compounds such as sulphides. The purifying properties of soil therefore largely depend on the capacities for oxygen renewal in its pores. Simple computations allow estimation of the quantity of oxygen contained in the pores of the top 30 cm of soil surface to be about 450 kg ha^{-1} when the soil is dry and decreasing to about 150 kg when its moisture reaches field capacity; gas diffusion enables recharging this gaseous atmosphere several times during the day. Moreover, many data indicate that soil may readily oxidise organic matter up to several hundred kg ha^{-1} day^{-1}. This oxidation changes considerably when the soil becomes water saturated. Nevertheless, if it is possible to artificially improve the conditions of oxygen transfer, as is done sometimes with sandy material, the oxidation capacity of the microflora including the requirements for nitrification, may exceed 10 t ha^{-1} day^{-1} of oxygen consumed.

- Temperature, together with oxygen, is the other physical parameter that determines the purifying capacity of the micorflora. It may generally be considered that metabolic activity, measured by respiration, increases twofold with an increase of 10°C in a temperature range of 10 to 25°C ($Q_{10} = 2$). Below 10°C, microbial activity slows down at a faster rate than may be anticipated by this Q_{10} of 2, without being completely stopped. It should be borne in mind, however, that except for a few centimetres from the surface, temperature variations in soils are considerably attenuated compared to air and deep down it tends to that of the remarkably stable aquifers at round 12-15°C.

- Another important characteristics of the purifying biomass is its association with the mineral matrix of the soil, which partly affects its function and explains why soil microflora is localised in the surface horizon and does not move deeper in the profile. This fixation can explain why metabolic activity of the microflora is proportional to the specific areas of the support (Fig. 13.5) and that in soils there is a positive correlation between the amount of microbial biomass and clay content. Fixation to the support also provides the microflora with high resistance to stress conditions.

During purification of water by soil particular attention is paid to nitrogen, some compounds of which are highly pollutant: ammonium in surface waters, nitrate in drinking waters and nitrous oxide (N_2O) in the

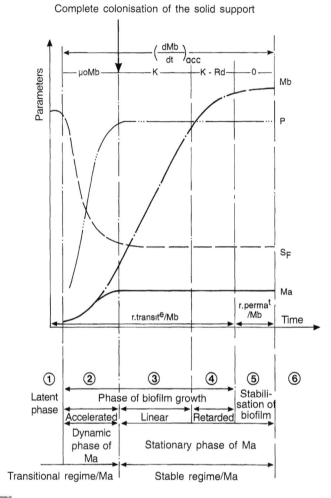

Complete colonisation of the solid support

F·ọ ١ϳ٦   Variations phases in development of a fixed bacterial culture (biofilm) on a mineral solid support in a renewed culture medium (Mb: total bacterial biomass fixed per unit surface on the support; Ma: active bacterial biomass on this same support; S: concentration of energy substrate at the outlet of the reactor; S_F: final concentration; P: concentration of products formed by the culture at the outlet of the reactor (R. Belkhadir, thesis, INSA, Toulouse, 1986).

atmosphere. Therefore, while treating nitrogenous products the aim is to maintain them in the soil in a non-dangerous and very slightly mobile form, or to ensure that they are absorbed by vegetation after their transformation into nitrate or even to return them to the atmosphere as inert nitrogen gas by bacterial denitrification.

Retention in the soil may be either in the form of ammonium that is effectively adsorbed on the clay-humus complex while remaining accessible to microbial transformations or by microbial immobilisation after synthesis of the cell constituents from ammonium or nitrate in the presence of a carbon substrate. Thus, ploughing back harvest residues, as done by farmers at the end of summer, helps to reduce leaching of several tens of kg ha^{-1} nitrate nitrogen towards the aquifer (Fig. 13.6). Lastly, denitrification may be desirable to eliminate excess nitrate nitrogen when it cannot be otherwise used. In cultivated soils that are adequately aerated, this anaerobic transformation may involve small quantities of nitrogen: a few kilograms per hectare. On the other hand, if soil with a high nitrate content can be temporarily placed in anaerobic conditions or water with a nitrate content is passed through such a medium, denitrification thus induced may lead to elimination of much larger quantities. Thus, in the literature there are reports of elimination of nitrogen to the extent of several hundred kg ha^{-1} in a few days and even several t y^{-1} in certain situations where purification is carried out. Nevertheless, for this purification treatment by denitrification, nitrogen must be nitrified earlier. This aerobic step is often a limiting one in the chain of transformations for soil-based treatment systems because of the several constraints with respect to inhibition of the nitrifying microflora (insufficient oxygen, too low temperature and/or pH, excess ammonia...).

Leached N (kg ha^{-1})

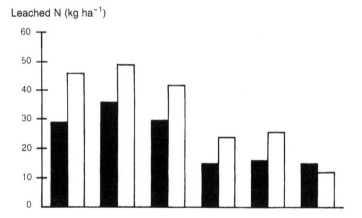

Fig. 13.6   Effect of ploughing back (■) or burning (□) grass straw on leaching of nitrates (in kg nitrogen per hectare) during winter following harvest in 6 different situations in England (Jarvis et al., 1989).

SOIL, A STORER OF FERTILISING ELEMENTS: RECYCLING BY VEGETATION

When the land-spreadings are on agricultural soil, the elements that can be assimilated by plants can also be recycled by the vegetation depending on its requirements. To limit transports to aquifers or accumulation in arable horizons, spreadings have to be adjusted according to possible exportations by vegetation, the quantity supplied being calculated from the first element for which the requirements are met by the supply. This theoretical step is actually used by agronomists for the principal fertilisers (N, P, K), taking into account their particular characteristics. Phosphates have limited mobility; the quantities supplied during a spread may be available over several years unless they are stably immobilised. Potassium can also be stored and is available over several agricultural cycles, but its assimilation varies depending on availability and may lead to considerable excess consumption; excessive amounts may cause imbalances in the mineral composition of the plants.

Supplies of nitrogen must take into account their biological availability, i.e., the capacity of the supplied products to be transformed to nitrates which cannot be stored in the soil profile. Simplified models are presently available which help to estimate the supply of nitrogen based on the characteristics of availability identified in a laboratory.[2] The supplies are than calculated so that the provision of mineral nitrogen from various available sources does not exceed the vegetation requirements. To be more complete, this approach would have to consider the mechanisms of losses that have to be compensated. This explains why these methods of forecasting are difficult to implement with precision.

For this agronomical approach to management of wastes to be practicable, it must be accepted that certain non-fertiliser products such as chlorides or sodium cannot be retained in soil and may migrate in the waters, the spread amounts being defined based on the quality criteria of the receiving aquifers. This agronomical management may sometimes be rendered completely impossible due to an accumulation of mineral elements, especially the previously mentioned heavy metals which cannot be easily extracted by vegetation. The objective of the manager of such spreads is to reduce supplies while accepting a minimum threshold of contamination, depending on the natural content of the soil which

---

[2]Muller and Dutil (Dutil, 1984) have proposed a model of the type: $N = N_0 + N_1 (1 - \exp(-k_1 t)) + N_2 (1 - \exp(-k_2 t))$, where N is the quantity of nitrogen supplied in time t from a provided quantity; N_0 is the mineral nitrogen provided with the waste; N_1 and N_2 the quantities of nitrogen defined as readily and slowly mineralisable; k_1 and k_2 the mineralisation rate constants.

should have no harmful effect on the quality of plant production or the quality of the percolating water.

Lastly, some situations in spreading appear easier to manage when there is no vegetation. The management of such situations should, even more than in the previously mentioned cases, take into account the phenomena of accumulation that appear and which, in time, will have repercussions on the quality of the percolating water.

SOIL, A PURFIER OF PATHOGENIC MICRO-ORGANISMS

Domestic and urban wastes and effluents from animal farms spread on soils contains pathogenic micro-organisms that can be transmitted to man and animals: helminths, protozoa, fungi, bacteria, viruses. These micro-organisms face unfavourable conditions for survival in the soil, which lead to their progressive destruction. Nevertheless, if their destruction is partial only, these pathogens may contaminate water or food products and maintain focus of parasitism in an endemic state, or cause more acute epidemics such as the one which occurred in Mexico in the 1960s and resulted in several thousand deaths.

Treating pathogenic micro-organisms with soil depends on three types of phenomena: their retention by the soil, their changed state in the soil and their transport in drainage waters or by runoffs. Their retention is due to filtration, as mentioned earlier, which is effective for the largest of them such as helminths and protozoa, and adsorption, which is a determining factor for bacteria and virus survival. The effectiveness of adsorption increases with clay content, the cation exchange capacity and the specific surface of the materials, and decreases with the moisture content of the soil. Some authors (Reddy et al., 1981) consider that the retention of micro-organisms by soil follows the same laws as the adsorption of chemical compounds and may be modelled in the same way by equations of the type

$$MRS = K \: MSUS$$

in which MRS is the microflora retained on the soil (number of micro-organisms per gram of soil), MSUS the microflora in suspension (number of micro-organisms per ml solution), and K the retention coefficient (ml g^{-1}), which depends on the particular characteristics of the micro-organisms and the soil. Adsorption is considered a determining parameter that regulates the evolution of viruses.

The evolution of pathogenic organism populations in soil depends on their capacity to develop or survive under the conditions of stress to which they are subjected. The micro-organisms maintained on the soil surface are exposed to the biocidal effect of ultraviolet rays from sunlight. In soil of temperate areas, they are subjected to temperatures that vary between negative values during periods of frost to a maximum that may exceed 30°C, with an average value of 15°C, far from their optimum, which is generally about 37°C. While their growth, often dependent on the host they parasitise, occurs in aqueous environments with a high content of readily biotransformable organic matter, in the soil they are often in oligotrophic, nutrient-poor environments and may be exposed to desiccation of fairly long duration and latter to osmotic pressures very different from those of the environments favourable to them. They are often subjected to competition for nutrients from numerous indigenous microflora generally well adapted to various milieus. Lastly, like the bacterial and fungal microflora in soil, pathological micro-organisms are subjected to predation by protozoa. Table 13.4 illustrates these effects of competition for nutrients and predation from the indigenous microflora on a population of potentially pathogenic bacteria.

Table 13.4 *Change in population of a pathogenic* Escherichia coli *introduced with waste water on two columns of sand 1.5 m in height, after passing through these columns priorly sterilised (treatment 1), and inoculated with a pure bacterial flora (treatment 2: indication of competition for nutrients) or inoculated with a mixed bacterial-protozoan flora (treatment 3: indication of the cumulative effect of competition for nutrients and predation) (after O. lung's thesis, Univ. Montpellier, 1993)*

Initial population	7×10^6 ml^{-1}
Population after treatment 1	5.5×10^6 ml^{-1}
Population after treatment 2	7.5×10^4 ml^{-1}
Population after treatment 3	7×10^3 ml^{-1}

Due to all these conditions, populations of pathogenic micro-organisms in the soil tend to reduce according to the known kinetics of the exponential type:

$$M_t = M_0 \exp(-kt)$$

In this relation M_0 and M_t are particular microbial populations at the time of supply (t_0) and at time t, and k is the rate constant of mortality (in day^{-1}). Measurement of this rate constant helps under particular conditions to calculate the half-life of these populations. The values observed for many bacteria and viruses in soil conditions, somewhat like those in France, vary from about 10 to 100 hours and even more. The effect of various factors of the environment on this constant k, can also be identified experimentally, which leads to the elaboration of models used to forecast the evolution of these micro-organisms in the soil. These models, based on a statistical approach, cannot be generalised without experimental verification, especially since a description of this evolution does not effectively take into account the ability of some of them to develop forms of slow life and resistance to unfavourable conditions of the environment, such as spores, cysts, eggs. In these resistant forms, micro-organisms await the return of more favourable conditions for their development. This also explains the considerable diversity of data in the literature on the survival of pathogenic organisms in soil, which varies from a few tens of hours for some viruses or bacteria to several months or several years for others.

Risks of being carried away from areas of spreads result from the aforementioned mechanisms and the hydraulic activity of the soil. These risks are higher in milieus where survival is encouraged, such as soil with a high water level. Risks of leaching are limited with an increase in thickness of the material that has to be passed through. Most of the work carried out with urban waste waters spread over predominantly sandy material reveals that the quantity of bacteria carried to depth increases with the quantities supplied, and reductions of 10^2 to 10^3 are usual for 1 to 1.5 m thickness of material to be passed through. This signifies that such a filtration eliminates 99 to 99.9% of faecal micro-organisms contained in the spread water. However, if the water initially contained 10^6 to 10^7 organisms ml^{-1}, after such a filtration it still contains 10^3 to 10^5 organisms ml^{-1}! Transport by runoff, i.e., unfiltered waters which flow on the surface, is observed mainly during periods following spreads and is the subject of particular attention in animal husbandry zones.

GEOPURIFICATION IN FRANCE

The purifying properties of soil, or geopurification, are used for treating various effluents: domestic and urban waste waters, industrial waste waters and effluents from animal farms.

Spreads of domestic waste waters through networks of underground drains originating from septic tanks are still common in France. Their management is usually somewhat poor and has led to the establishment of collective treatment networks. On the other hand, large-scale spreads of urban wastes are not common. A few areas for such spreads still exist in Achères, west of Paris, which have been retained to supplement the existing treatment plants. There are several small installations for spreads of untreated waters in different regions which are used for demonstration. On the other hand, filtration of partly treated water, spread on soils or on porous sandy material and meant to replenish the water in aquifers used for drinking purposes or irrigation, has increased largely during the last 20 years. About 60 installations of this kind were surveyed in France in 1991. Recent work on management of these treatment plants receiving water levels of about 500 mm per day was intended to optimise oxygenation by improving air circulation, identifying conditions to limit clogging by alternating phases of activity and rest, and studying conditions for elimination of pathogenic germs.

Spread of waste waters from agricultural and food industries has become a common and accepted practice in the last three decades for wine cellars, canning factories, sugar industry and potato-starch production. The effluents from these industries generally contain organic matter which is readily biodegradable and the composition of mineral elements is similar to that of agricultural products, which helps in their use. The amounts used are calculated on the basis of the spread fertiliser elements, notably nitrogen, according to the earlier described models. The salt content may be an obstacle for such spreads: some manufacturing chains, such as sugar factories, have been able to eliminate the use of sodium chloride, especially for the regeneration of exchanger resins, while others continue to use this salt in particular for pickling, cheese-making and sauerkraut.

These spreads continue to be the traditional mode for treatment of effluents from animal farms. The amounts used are based on an estimate of the nitrogen requirements of the vegetation. Nevertheless, the increase in size of production units has led to excessive releases compared to the available areas and to the present situation with respect to pollution by nitrates and phosphates observed in all regions of the world where animal husbandry is intensive. Experiments are now underway to develop methods for treatment at these farms based, among other things, on intensification of the sanitising properties of soil to accelerate nitrification and denitrification processes and thereby ensure elimination of a large part of the nitrogen.

FURTHER READING

Anonyme. 1983. Protection des sols et devenir des déchets. Recherche et environnement, vol. 26, 430 pp.

Calvet R. 1990. Nitrates, agriculture, eau. Inra Éditions, Paris, 576 pp.

Chamayou H, Legros J.-P. 1989. Les bases physiques, chimiques et minéralogiques de la science du sol. Presses Universitaires de France, 594 pp.

Dutil P. 1984. Utilisation agricole des déchets urbains et industriels et problèmes de pollution liés à l'environnement. Techniques agricoles, 1390: 1-18.

Germon J.-C. 1985. Le sol, un système épurateur efficace s'il est bien géré. Revue du Palais de la Découverte, 14: 19-41.

Iung O., 1993. Épuration bactériologique en infiltration-percolation. These de doctorat, Université des Sciences et Techniques du Languedoc, Montpellier, 172 pp.

Lefevre F., 1988. Épuration des eaux usées urbaines par infiltration-percolation: étude expérimentale et définition de procédés. Thèse de doctorat, Université des Sciences et Techniques du Languedoc, Montpellier, 341 pp.

Reddy KR, Khaleel R, Overcash MR. 1981. Behavior and transport of microbial pathogens and indicator organisms in soils treated with organic wastes. J. of Environ. Quality, 10: 255-266.

Soil Quality Observatory: A Management Tool for Agriculture, an Instrument for Monitoring Ecological Systems

S. Martin

INITIAL OBJECTIVE

Soil undergoes slow, discrete and irreversible degradations. This is an insidious disorder. The economic consequences are serious. They penalise mainly the local inhabitants and farmers. The heritage bequeathed to subsequent generations is inexorably destroyed. The Soil Quality Observatory (SQO) was created by the Ministry of Environment in 1984 to obtain better knowledge of these phenomena, forecast their changes and encourage research to find solutions (INRA, 1983; Mamy, 1993; Martin, 1993; Martin et al., 1998).

PRESENT STATUS OF THE SQO

The SQO is based on a network of monitoring sites, each of about one hectare, selected for their representativeness. The soils and the vegetation on them are studied every five years: classic soil properties, heavy metals, radioactive elements (see Table A.1 below).

The various partners (Ministry of Environment, INRA, CNRS, universities, CEA, professional agricultural organisations, ...) and a scientific committee carry out the work of the SQO. A computerised database is gradually being established.

Table A 1 *Minimum programme for measurements and observations common to all sites of the SQO*

During site establishment	General description of site (historical, geographic context...) Detailed pedological study
Every five years	• Classic chemistry and physicochemistry: —granulometry (five fractions) —water pH —organic carbon —organic nitrogen —cation-exchange capacity —exchangeable calcium —exchangeable magnesium —exchangeable potassium —exchangeable sodium —phosphorus (P_2O_5) —total calcium carbonate ($CaCO_3$) • Trace-metal elements —cadmium —chromium —cobalt —copper —nickel —lead —zinc • Radioactive elements (gamma emitters) —specific radioactivities of nuclides —potassium 40 content —thorium 232 content —uranium 238 content
Continuously	• Events affecting site (rotation of crops, introduction and removal of polluting products...)

(Table Contd.)

(Table A.1 contd.)

Not yet taken into account for technical and scientific reasons, but soon to be included	• Biological properties —earthworm populations —microbes (biomass and activity)
	• Erosion
	• Physical properties of soils
	• Effects of pesticides
	• Quality of harvests

The SQO seeks essentially to visualise soil states. It also constitutes a stable link between fundamental research and serious concern for the environment, a network of soil laboratories capable of facilitating integrated research and providing coherert sets of data for modelling, and a tool for training agricultural professionals in new techniques for studying soil. It may be noted that due to its paritcular objectives, the SQO poses new issues for research.

FUTURE OF THE SQO

While closely dependent on the global socioeconomic context, the subjects discussed during the Summit on Planet Earth in June 1992 at Rio de Janeiro centred around three important questions recognised as priority matters with respect to the environment: the ecological aspects of planetary changes, ecology and preservation of biological diversity and strategies for sustainable ecological systems. A committed assessment of these major concerns in political, economic and technical decisions necessitates the availability of actualities for assessing changes in the biosphere as a consequence of human activity (Barbault, 1990; Lubchenco et al., 1991).

The SQO may be considered a special instrument for monitoring changes in the biosphere. Its field of intervention will have to be extended to relationships between the greenhouse effect and terrestrial ecosystems, biodiversity monitoring within benchmark terrestrial ecosystems, encouraging a sustainable agriculture....

As of now, the SQO has been designed mainly for monitoring changes in soils in a certain number of well-defined situations and 11 observation sites have been established (Fig. A.1)

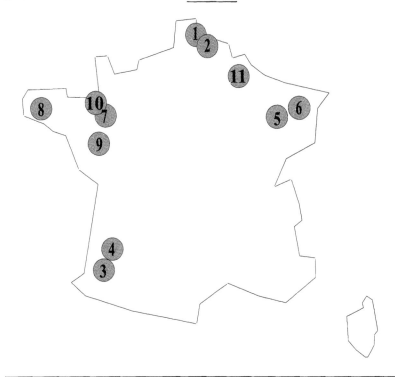

Department	Characteristics	Soil type
1 North	Large-scale farming	Brown Clay Fluviosol
2 Pas de Calais	Large-scale farming, soil highly contaminated by atmosphere fallouts (cadmium, lead, zinc)	Redoxic Neoluvisol
3 Landes	Forest	Duric Podzosol
4 Landes	Large-scale farming	Resaturated Brunisol
5 Vosges	Large-scale farming	Pseudoluvisol
6 Vosges	Forest	Typical Allocrisol
7 Ille et Vilaine	Forest	Typical Allocrisol
8 Finistère	Large-scale farming	Oligosaturated Brunisol
9 Loire (Atlantic)	Large-scale farming	Mesosaturated Brunisol
10 Channel region	Market gardening in polders	Calcareous Thalassosol
11 Ardennes	Large-scale farming. Spreading of sludges	Broken Neoluvisol

Fig. A 1   SQO sites in 1998.

It is intended to expand the SQO in three directions (Legros and Martin, 1997; Martin et al., 1998):

- Geographical extension at three levels with:
  - several sites better equipped for a study of processes (for example: with measurement of flux of pollutants);
  - a network of permanent sites (several tens) broadly comparable with the present sites in order to assess the essential aspects in situations existing in France;
  - a network of minor investigation sites (a few hundred) to estimate the spatial extent of changes that have been studied in local situations, to ensure better locations for permanent sites and to establish a warning system.
- Thematic development. The present programme for the monitoring of sites will soon be completed by a study of some biological properties of soils. Other aspects will have to be examined as soon as possible: assessment of the effects of pesitcides and other specific organic molecules, monitoring of the physical states of soils and erosion, precise estimate of the quality of harvests.
- Strengthening of links with the stakeholders and in particular closer collaboration with professional agricultural organisations, diversification in applications of the database and publication of thematic documents.

The concept and implementation of a network for monitoring soil quality is a multidisciplinary effort combining daily management, technical innovation and scientific research. The problems are many and complex. There is (1) the immediate requirement of associating additional partners and producing reliable information and (2) the longer term necessity of keeping aware of stakeholder needs and of integrating scientific advances.

SQO: EXAMPLE OF SCIENTIFIC PROCESSES APPLIED TO THE TERRAIN

Monitoring an Ecosystem Means Asking it Questions

Questions determine choice of measurements and observations. The manner of setting a problem depends on the type of visualisation of the ecosystems one wants.

For the SQO the main objective is to detect and evaluate changes in the soils of France. Hence the SQO sites are selected according to:

- type of soil,
- type of land use (large-scale farming, grassland, forests, ...) and
- type of soil degradation (Baize, 1988).

The minimum programme for measurements and observations implemented on sites of the SQO has been defined taking into account:

- state of knowledge about the structure and functioning of soils,
- what is *a priori* known about the problems of soil degradation,
- technical feasibility for measurement and observation,
- scientific ability to interpret the result,
- financial constraints.

Obviously, completion of the programme depends on the specific nature of each site.

The questions vary depending on the categories of stakeholders in an observation system. In the case of the SQO it may be broadly stated that scientists mainly use detailed data that have been subjected to a single validation, while agricultural professionals need data that have been interpreted and which they can rapidly integrate into their own soil-management systems (for example: forecasting effects of a particular agricultural practice on soils, helps in assessing the quality of soils in a particular plot...), while political decision-makers most often require synthesised information which provides global perspectives (for example: effects of a particular economic activity on soils; the main degradations in soils over a specific geographic area...). The SQO must therefore serve several objectives while retaining its own coherence.

Observation Scales

Questions determine scales of observation and the period between two observations: ecosystems appear different in different scales of observation. Thus, mechanisms for fixation of metal pollutants by soils are studied at the microscopic scale (a few μm to several mm). Phenomena of transport of metal pollutants from soils to plants are generally observed at the scale of a square metre, on plants kept in pots. A hectare, the area of SQO sites, appears to be highly suitable for monitoring of soil quality; it is the level of the agricultural parcel. A change to the level of the catchment area is necessary for studying the relationship between management of soils and the quality of groundwater.

The problem of relating these various scales arises unavoidably when a natural system is observed. The sampling strategy for the SQO links the scale of soil sample with that of the agricultural plot or, more precisely, enables estimation of certain properties of a parcel based on measurements taken on the samples. Further, if the SQO sites are suitably selected, it may be expected that they will give an overview of principal issues at the national level that corresponds to a new change of scale (Frontier, 1983; Leprêtre and Martin, 1994; ORSTOM, 1990).

Significant Changes, Natural Fluctuations and Heterogeneity

The major problem is to recognise and measure the responses of ecosystems to stresses caused by human activity (introduction of pollutants, changes in landscape, introduction or suppression of animal or plant species...). A lack of sensitivity of indicators to environmental stresses limits detection of early stages in changes and consequently the possibility of managing ecosystems.

As undisturbed systems may vary in time, it is essential to take into account the natural variability over time in the descriptors of the physical environment and the selected biological indicators in order to detect undesirable changes. The best indicators maximise the ratio 'sensitivity over variability in time' (see Figs. A.2 and A.3).

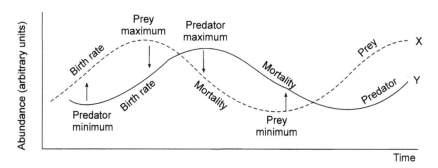

Fig. A. 2 Natural variability in time: oscillations in different phases of abundance (or densities) of prey and predators. The maximal (or minimal) rate of increase in predator population occurs at the time of maximum (respectively minimum) prey density. The rate of multiplication of the predator population depends mainly on the availability of food. The oscillations are continuous (after Frontier and Pichod-Viale, 1991).

Moreover, the properties of soils (contents of micropollutant, organic matter, microbial biomass, ...) are not generally distributed in a uniform manner in space. When two measurements taken after an interval of several years provide different results, it is necessary to know whether this variation can be attributed to the heterogeneity of the particular plot and the variability of the measurement itself or actually signifies a

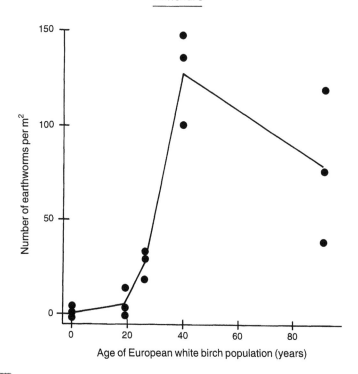

Fig A 3 Ecological response to stress: change in density of all species of earthworms in a heath with common heather (*Calluna vulgaris*) after colonisation with European white birch (*Betula pendula*) depending on age of the birch population. Colonisation of the heath by white birch causes a severe decrease in common heath, a change in the soil (increase in pH, exchangeable calcium, total phosphorus, mineralisatin of nitrogen and decomposition of cellulose) and a very pronounced increase in biological activity in the soil. These phenomena are reversed when the European white birch population grows old and the trees die, enabling the common heather to reinvade the site (after Miles, 1985).

change in the soil. This question is central to the problems that arise during elaboration of the SQO sampling strategy, which helps to explore different components of variability in the results: internal, in an analytical laboratory and spatial, both global and local (see Figs. A.4, A.5, and A.6).

Interdisciplinarity must be Organised

The SQO makes use of scientific and technical disciplines as diverse as pedology, biochemistry, microbiology, soil physics, plant physiology, ecology, agronomy, chemistry, physics of radioelements, statistics, data

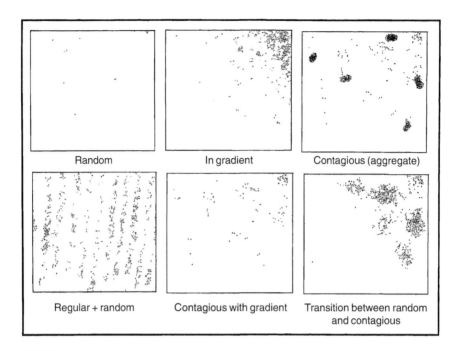

| Random | In gradient | Contagious (aggregate) |
| Regular + random | Contagious with gradient | Transition between random and contagious |

F.g 4 4   Spatial distributions of density of an element (after Leprêtre and Martin, 1994).

processing, climatology, hydrology, geodesy.... Each of these disciplines, in relation to the soil and within the general framework of the SQO, has its own requirements, its own levels of approach, methods of investigation and interpretation. Interdisciplinarity does not occur spontaneously but, on the contrary, must be organised. This involves interrelating the various objects which constitute the different disciplines. The tools available are databases, simulation models and artificial intelligence.

The computerised database of the SQO is an implicit translation of most of the SQO's objectives and the choices consequently made. It provides various participants of the SQO with a common language and a means for rapid communication among themselves, without error in interpretation and at the least cost. A significant example: identification of each soil sample. It was decided that each soil sample taken from SQO sites would be identified by the following sequence of information:

[Number of the SQO site/date of sampling/number of the point of sampling in the sampling plan used on the date of sampling/depth of sampling (upper and lower limits).]

Samples \ Stations	1	2	3	4	5	6	7	8	9	10	11	12	13
a	×	×	×		×			×			×	×	
b	×	×		×		×					×		
c	×	×			×			×		×			
d	×	×				××× ××× ×××	××× ××× ×××			× × ×	×		

× × × Replicates of core sample

(Operative) variance: 3 selected samples

Local variance (stations): 4 selected stations

Global variance (site): 9 selected stations

Fig. A 5 Selection of samples from an SQO site for the first phase of analysis. Twenty-nine samples are taken from 29 extractions, with 3 samples providing 3 extractions each to enable evaluation of variability of the results associated with internal causes in the laboratory (operative variability). If necessary, in the second phase, the 29 remaining extractions may be subjected to analysis. This step in two phases is for reasons of economy.

Elaboration of this rule, common for all participants in the SQO, was even more necessary because to study the relations between various soil properties, each soil sample is usually examined from the viewpoint of each of the different disciplines.

Further, precise formulation of phenomena requires simulation models and applications of artificial intelligence, which are valuable aids in interdisciplinarity. For its part, the SQO is likely to provide coherent sets of data for modelling (for example: computer simulation of the flow of trace-metal elements through a catchment area and variations in the reserve of these trace elements) and/or artificial intelligence (for example: expert system assisting agricultural professionals in conducting diagnoses of soil quality).

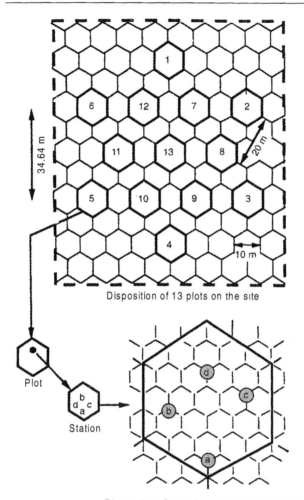

Disposition of 13 plots on the site

Disposition of sampling points at each station

Fig. A.6 Sampling plan on SQO sites. The sites (area # 1 hectare) are actually divided into plots. A station (area # 3.5 m²) is selected within each of these 13 plots (the tops of two concentric hexagons + the centre). Four samples are taken from each station which makes 52 samples in all.

REFERENCES

Baize D. 1988. Méthodologie relative au choix, au prélèvement et à l'étude pédologique préalable des sites de l'Observatoire de la Qualité des Sols. In: Manuel de l'OQS. Ministère de l'Environnement, Paris, France.

Barbault R. 1990. Écologie générale. Structure et fonctionnement de la biosphere. Masson, Paris, 269 pp.

Frontier S. (ed.). 1983. Stratégies d'échantillonnage en écologie. Masson, Paris, 494 pp.

Frontier S, Pichod-Viale D. 1991. Écosystèmes: structure, fonctionnement, évolution. Masson, Paris, 392 pp.

Institut National de la Recherche Agronomique. 1983. Étude de faisabilité d'un Observatoire de la Qualité des Sols. Ministère de l'Environnement, Paris: 112 pp.

Legros JP, Martin S. 1997. L'Observatoire de la Qualité des Sols à l'aube d'un nouveau développement. In: Le sol, un patrimoine à préserver Chambres d'Agriculture, France, Special Publication, 886: 45.

Leprêtre A, Martin S. 1994. Sampling strategy of soil quality. Analysis Magazine, 23(3): 40-43.

Lubchenco J. et al. 1991. The Sustainable Biosphere Initiative: an Ecological Research Agenda. Ecology. Ecological Society of America. April 1991, 72(2): 371-412.

Mamy J. 1993. Qualité, usages et fonctions des sols. In: La qualité des sols. Chambres d'Agriculture. Special Publ. 817, pp. 6-7, France.

Martin S. 1993. The "Observatoire de la Qualtié des Sols", an example of ecosystem monitoring. In: Integrated Soil and Sediment Research: A Basis for Proper Protection. Heijsackers HJP and Hamers T (eds.). Kluwer Acad. Publishers, Dordrecht, Netherlands.

Martin S, Baize D, Bonneau M, Chausood R, Gaultier JP, Lavelle P, Legros JP, Leprêtre A., Sterckeman T. 1998. The French National "Soil Quality Observatory" 16th World Congress of Soil Science, symposium 25. Montpellier, France, August 1998.

Miles J. 1985. Soil in the ecosystem. In: Ecological interactions in soil: plants. microbes and animals. Fitter AH (ed.). Special Publ. n° 4 of the British Ecol. Soc. Blackwell Scientific Publications, Oxford, pp. 407-427.

Orstom Editions. 1990. Le transfert d'échelle. SEMINFOR 4, Paris, 517 pp.

For Product Safety Concerns and Information please contact our EU
representative GPSR@taylorandfrancis.com
Taylor & Francis Verlag GmbH, Kaufingerstraße 24, 80331 München, Germany